First Edition
Rev. Fall 2022

Librande

# Human Anatomy Laboratory Manual

Human Anatomy Laboratory Manual by Jason Librande

Published August 2022
Los Angeles, CA

# Table of Contents

# For Instructors

Dear Instructor,

Thank you for taking the time to consider adopting this book for your course. As you know, very few students are affluent enough to easily afford the expensive textbooks we assign them each semester. Especially in the context of laboratory manuals, it is unlikely that they will be able to resell these for any money; writing on pages, and worse, tearing them out, disqualifies them from being sold used.

As I wrote this laboratory manual, I focused on two guiding principles:

1. Create a product that rivals the efforts of mainstream publishers
2. Create a product that resonates with our students in terms of diversity, equity, inclusion, and anti-racism.

In regards to the former, you will find hundreds of openly-licensed images in this lab manual. Many of these were heavily modified versions of the work of others. The intent is to follow the labeling heavy design of mainstream publisher lab manuals. The benefit of this is that this lab manual is not limited by your resources. Any institution can adopt this lab manual with ease. Likewise, I have developed ancillaries that you may find helpful. This includes files of scanned microscope slides, rotatable photos of anatomical models, and copies of images that were created specifically for this lab manual. For more information, see jlaandp.com. My hope is that all of this will make adoption of this OER lab manual easy. Through this, we can greatly help our students.

One of the benefits of OER is the ability to customize it to your needs. Part of my own exploration of OER has been crafting resources that reflect the diverse student body that I teach. We are all well aware of the fact that students do better when they feel part of the class. To accomplish this second principle, I have included call-outs called "Every-body is Different". These tell stories and information relevant to a student body with high BIPOC enrollment. Of course, there are numerous other ways to increase diversity in the material. Images, for example, have been carefully chosen to represent a variety of races. The non-European history of anatomy is also addressed. Despite these efforts, my students are not your students. If you are utilizing a PDF version of this lab manual, I encourage you to visit jlaandp.com to download the editable word documents. You will then be able to craft your own lab manual that incorporates the cultures, histories, and diversity of your study body.

I hope that you find the contents of this lab manual useful for your teaching. It will also be exciting to see what branches spring out from sapling; I am sure the materials you will create will outdo what I have here.

Jason Librande

*For Donnie, Helga, and Lucky. A dedication is probably the last thing you want as you do not read English, and paper is not edible.*

# Lab Rules

These lab rules are not meant to supercede your instructor's. If they provide their own lab rules, follow those.

1. **No food or drink!** Even though much of what we will do resolves around (non-toxic) chemicals, this is still technically a lab room. Other classes may use dangerous chemicals, and perform other experiments in this room.
2. **The way you come into the room for the day is the way it should be when you leave.**
3. **Spray down benches before you begin, and after you finish**
4. **Put things back properly!** This includes microscopes, chairs, models, etc. We may very well only have one of an item (e.g., certain slides), if we do not care for it we cannot purchase a new one. As you might expect these items are very expensive. A new, high quality microscope costs well into the thousands!
5. **No rough-housing, sitting on tables, "fooling around" etc.**
6. **No guests, children, pets, etc.** Many times we will be dealing with chemicals that are hazardous, and any guests are a liability.
7. **Wear sensible clothing**. This generally means **closed-toed shoes, no dangling pieces of fabric, minimal fragrances, hair tied back, acrylic nails removed, and open wounds or sores should be covered**.
8. **Wear any protective equipment your instructor recommends.**
9. **Know where the following safety items and disposal receptacles are:**
    - Eyewash and shower
    - Fire extinguisher
    - Emergency phone
    - Regular trash can
    - Biohazard and specimen disposal can
    - Classroom exits

8. **Notify your instructor as soon as possible of an emergency or issue.**
9. **Even if you didn't make the mess, cause the issue, or break the item you still have some responsibility for it.** We all share the resources of this lab and college, please ensure that other students can utilize these resources as well. We appreciate any help in keeping our lab room organized and functioning.
10. **If you have any questions, ask!** Never guess. Your instructor is more than happy to answer them.

# Rules Agreement Sheet

By signing below, you agree that you understand the rules of the course, and agree to abide by them.

Failure to abide by these rules can potentially result in any disciplinary action ranging from deduction of points to a failing grade in the course with a referral to the disciplinary board.

Name (Printed):____________________________

Signature:____________________________ Date:_______________

If you are under 18, please take this form to your guardian and have them complete the portion below.

I, __________________ declare that I am the parent/guardian of the student identified in this form. I acknowledge that I have read the prior page.

Signature:____________________________ Date:_______________

# Lab 1

# Introduction to Anatomy



**Learning Objectives: Students will be able to...**

- Describe, briefly, the history of anatomy
- Use root words and rules to define and manipulate anatomical terms
- Define and use all the common anatomical directions
- Relate body parts to each other using anatomical terms
- Use anatomical terms to describe various parts of your body

## <u>Activity 1:</u> A brief history of anatomy

Anatomy is perhaps one of the oldest subfields of biology. Its roots extend back to ancient Greece where Aristotle (384-322 BC) coined the term anatome which means "to cut up", of course relating to dissection. This was the beginning, but likely even as hunter-gatherers we were already constructing an understanding of anatomy. Of the 1,600 or so years after Aristotle, much (but not all) of anatomy stuck close to the Mediterranean region. Many famous anatomists such as Galen (AD 150-200) and Hippocrates (noted for the famous Hippocratic oath) were born in Greece. Similarly, as the Renaissance came, Italian artists such as Michelangelo and DiVinci developed a great interest in the human form, including dissecting cadavers in order to draw them in exquisite detail. Other Italians such as Malpighi (AD 1628-1694) took academic posts at universities, but they were also excellent artists. Even non-Italians like Vesalius (AD 1514-1564) born in Brussels, quickly made their way to Italy to explore Anatomy.

From there, new technologies such as the telegram and steam-powered engines allowed for the spread of anatomical knowledge worldwide. Countries that had small roles in anatomy up to this point, such as England and the United States, picked up on this knowledge and integrated this information into their curricula.

It is important to note that up until the last century or so, all knowledge of anatomy was contained in drawings, not photos. This is one reason why artists such as DiVinci were able to leave such large impact on the field of anatomy despite not being considered "anatomists". Their skill with art allowed them to communicate information. On the following pages are some important drawings from the early study of anatomy.

A likely self-portrait of the famous Italian anatomist Andreas Vesalius (AD 1514-1564)

Vesalius' most famous book was *De humani corporis fabrica* (*On the Fabric of the Human Body*). It was one of the most important books written on the subject and helped to spearhead a renewal in the study of anatomy in Europe.

While art influenced how anatomy was understood and communicated, the relationship was reciprocal. For example, notice the kidney shape found within Michelangelo's (AD 1470-1564) fresco at the Sistine Chapel. Of course, to draw the human form correctly one must understand structures underlying the skin. This is key to why many Italian artists created such life-like and realistic paintings.

Perhaps one of the most famous "anatomical" drawings is DiVinci's (AD 1452-1519) "Vitruvian Man". While often co-opted for the teaching anatomy, the drawing itself has little to do with anatomy and moreso with mathematics. The image is meant to depict the "ideal" proportions of the human body as dictated by the Italian architect Marcus Vitruvius Polio.

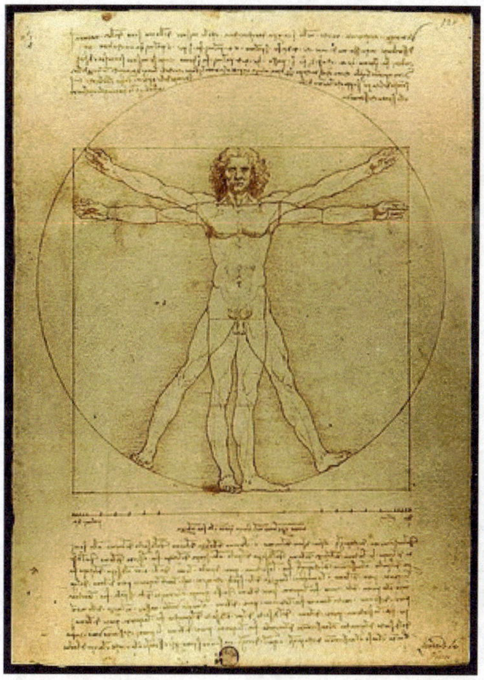

1

One of the biggest misconceptions regarding anatomy is that only Europeans have contributed to the field. That can't be further from the truth. As long as humans have been around, we have been curious about how we are structured. The reason why anatomy is often tied back to Europe is due to the fact that the modern systems of higher education are based off of the Medieval European system from hundreds of years ago. It is natural that these medieval universities had Euro-centric foci; knowledge could not spread quickly during that time. Unfortunately, the side effect is that Euro-centricity has been embedded in the modern university. Hundreds of years of Euro-centricity is hard to shake.

That is not the way things should be though! We have the internet now, so information travels quickly. We can share cultural histories of anatomy that may have been ignored by the western academy. In turn, our understanding of anatomy is becoming more diverse and multicultural. Here are some major contributions by anatomists of non-European decent:

A statue of Sushruta (~600 BCE, India). He is considered the Father of Surgery in part due to his book *Sushruta Samhita.* It provided detailed accounts of surgery long before any Italian anatomists began their work.

Cardiocentric theory, the idea that the heart is the seat of emotions and thoughts, is thought to originate from Ancient Egypt. It was believed that Egyptian Gods would weigh a deceased person's heart against a feather to determine if it was heavy (with guilt). The idea that the heart contains are emotion is still tied to the use of hearts to represent love.

Charles Drew (1904-1950) was a Black, American Surgeon. His work on blood transfusion and storage aided the U.S. greatly during World War II. Of great importance, is that he laid the groundwork for the desegregation of blood donation. Charles Drew University is named in his honor.

[2] Public Domain

## Review Questions:

1. What two European countries had a large impact on the study of anatomy?

2. Was Europe the only continent that contibued to anatomy? What other countried played a role?

3. What other skills did anatomists have to master other than being good scientists?

4. What are the names of three important figures in the study of anatomy? (there are many more than just three!)

## Application Questions:

1. What two languages do you think many anatomical terms are based on (hint: remember we are talking about "ancient languages", not modern-day languages)

2. Using the internet, find an anatomist who is not from a Europen country, the U.S., or Canada. Where are they from? When did they live? What did they research/discover?

## Activity 2: The language of anatomy

With a subject such as math or history, miscommunications are generally forgiving. Anatomy is much different. How you speak, and the accuracy of your spelling could have drastic effects on someone's health.

1

Consider someone who is having their eye removed due to retinoblastoma (a type of rare eye cancer). On their chart you write the shorthand O.S. (left eye) as the one to be operated on. However, you actually meant to write O.D. (right eye)! A surgeon in a rush may not notice this mistake and start the operation. Even though this is just a simple one letter mistake, it can result in a huge impact on someone's health. This is called a wrong-site procedure and it occurs approximately every 1 in 113,000 surgeries. But when considering the entire U.S. that is about 400 cases per year.

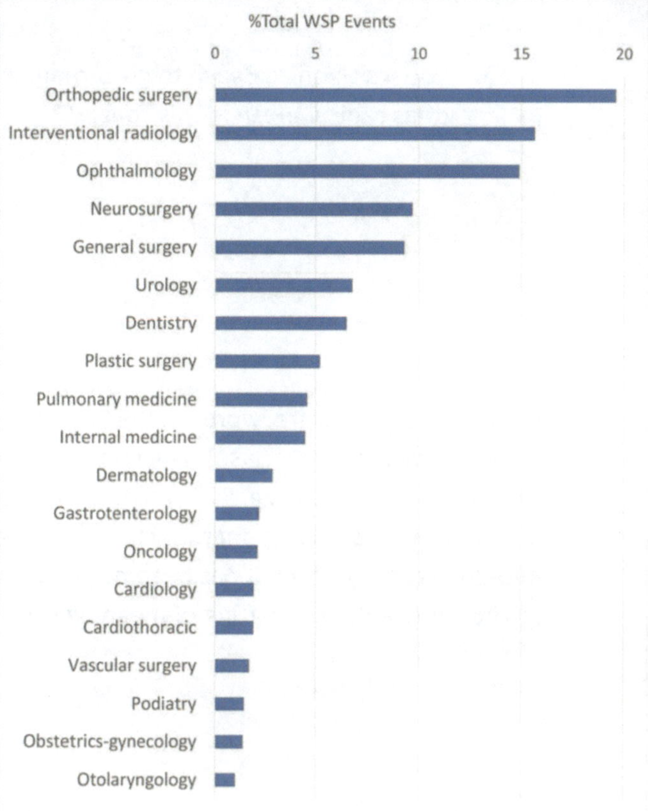


Certain specialties are more prone than other, such as those that require differentiating between left and right (e.g., ophthalmology).

In turn, spelling is critical to be successful in Anatomy, and the healthcare field as a whole. See your instructor for their policy on correct spelling for assignments and exams.

Another unique aspect of anatomy is that it must be communicated universally. Health care practitioners from around the world, and from many different institutions must be able to communicate harmoniously. As with other sciences, the language of choice is Latin. It is a particularly useful language because is considered a "dead" one. It is no one's native language, and outside of high-ranking Catholic officials, no one uses it conversationally. Consider the usefulness of a dead language by completing the questions on the next page:

[3] "Wrong-Site Procedures: Preventable Never Events that Continue to Happen" by Andrew Lin, Brian Wernick, Julia C. Tolentino and Stanislaw P. Stawicki is licensed under CC BY 3.0

## Try these questions:

1

1. You attend the first performance of Stravinsky's *Rite of Spring* in 1913 (which resulted in a riot). You proclaim the ballet as "radical". What does radical mean in this context?

2. You visit Venice Beach to go surfing, your friend describes your surfing ability as "radical". What does radical mean in this context?

As you are likely aware, words in the English language tend to morph as time goes on. As people use these words there can be subtle shifts in their meaning until 100+ years later the words have a completely new meaning. Even anatomy is subject to this. "Gross Anatomy" does not mean disgusting anatomy, it means large anatomy (i.e., study of large structures like muscles). Gross became slang in the mid-20th century where it has become synonymous with ugly or disgusting. Using a dead language such as Latin prevents this shifting of meaning since so few people use it.

## Activity 3: Using and manipulating anatomical language

English is derived, in part, from Latin. As speakers of English you already know, or at least have a guess as to what certain words mean. For example...

### Try these questions:

1. Consider the words microscope, microanatomy, and microsurgery. The prefix micro- means what?

2. Without knowing what microanatomy is, make a guess about what it is using the Latin prefix micro- to help you.

The same logic as above permeates all of anatomy. This is the idea of **etymology**; The history of a word and its "literal meaning". If you know your Latin, you can make a pretty good guess about the meaning (and sometimes even the location) of many structures. You can do this by simply splitting words up into roots, prefixes and suffixes to make an educated guess about the word's meaning. You will practice this at the end of this activity.

There are a variety of rules that come with the use of Latin, but in particular, you will need to pay attention to these rules for changing singular words to plural ones. Unlike English where 's' can usually be added to create the plural form, words from Latin origins are trickier:

**Table 1. Modifying Word Endings to Create Plural Words**

| Word ending | Remove the following | Add the following | Example |
|---|---|---|---|
| **-a** | | -e | larva becomes larvae |
| **-ax** | -ax | -aces | thorax becomes thoraces |
| **-ex or -ix** | -ex or -ix | -ices | cortex becomes cortices |
| **-ma** | | -ta | stoma becomes stomata |
| **-is** | -is | -es | anastomosis becomes anastomoses |
| **-nx** | -x | -ges | larynx becomes larynges |
| **-on** | -on | -ia | ganglion becomes ganglia |
| **-us** | -us | -i | nucleus becomes nuclei |
| **-um** | -um | -a | ischium becomes ischia |
| **-y** | -y | -ies | biopsy becomes biopsies |

While Latin is by far the most dominant language of anatomy, there is also some Greek as well. In particular, memorize the following number prefixes:

**Table 2. Latin and Greek Number Prefixes**

1

| # | Latin | Greek |
|---|---|---|
| *1* | unus | mono |
| *2* | duo | di |
| *3* | tres, tria | tri |
| *4* | quattuor | tetra |
| *5* | quinque | penta |
| *6* | sex | hexa |
| *7* | septem | septem |
| *8* | octo | octa |
| *9* | novem | ennea |
| *10* | decem | deca |
| *11* | undecim | |
| *12* | duodecim | |

## Review Questions:

1. Using Appendix 1 in the back of your lab manual, give the literal translation/definition for the following words:
   a. Cnidocyte:______________________________
   b. Microvilli:______________________________
   c. Tympanic:______________________________
   d. Phagocyte:______________________________
   e. Pericarditis:______________________________
   f. Hypochondriac:______________________________
   g. Myocyte:______________________________
   h. Menstruation:______________________________
   i. Ligament:______________________________
   j. Statolith:______________________________



2. The following table contains a variety of singular or plural forms of words. Use the rules on the previous page to convert between singular and plural forms and fill in the blanks.

| Singular | Plural |
|---|---|
| | Meninges |
| Phalanx | |
| Cerebrum | |
| | Neuroglia |
| Humerus | |

## Application Questions:

1. How long do you think the duodenum of the small intestine is? (see Table 2 for help)

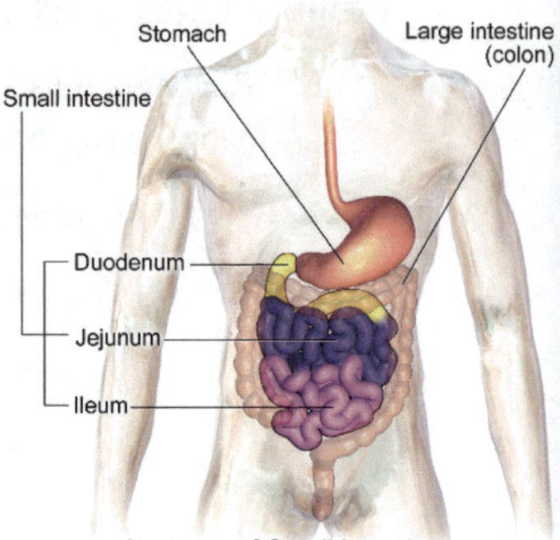

Anatomy of Small Intestine

2. What is the plural form of octopus? (read the next section to check your answer)

If you said **'octopi'** you are not necessarily wrong Octopus is an odd word to turn into plural form. The prefix octo- is of Greek origin (the Latin language stole it from Greek), but the suffix -us is of Latin origin. This results in a word that is both Greek and Latin. According to our rules as mentioned in Table 1, octopus should become **octopi**. This is perhaps the most common plural form of octopus; Treating it as a completely Latin word. But, being a partially Greek word, some grammatists believe it should be **octopodes** (oc-top-o-dez), a Greek plural form for a partially Greek word.

What trumps all of this is that we are speaking American English, not Greek or Latin. When a word such as octopus enters our language it adapts to our rules. This means that **octopuses** is the actual "correct" plural form of octopus. Correct being in quotes because English is a language where there is great flexibility in the "rules".

---

[4] "Small Intestine" by BruceBlaus is licensed under CC BY 3.0

## Activity 4: Anatomical terms and directions

In addition to the above Latin, there is also a variety of anatomical terms used to describe regions of the human body. As with Latin, knowing these words will be critical for your study of anatomy.

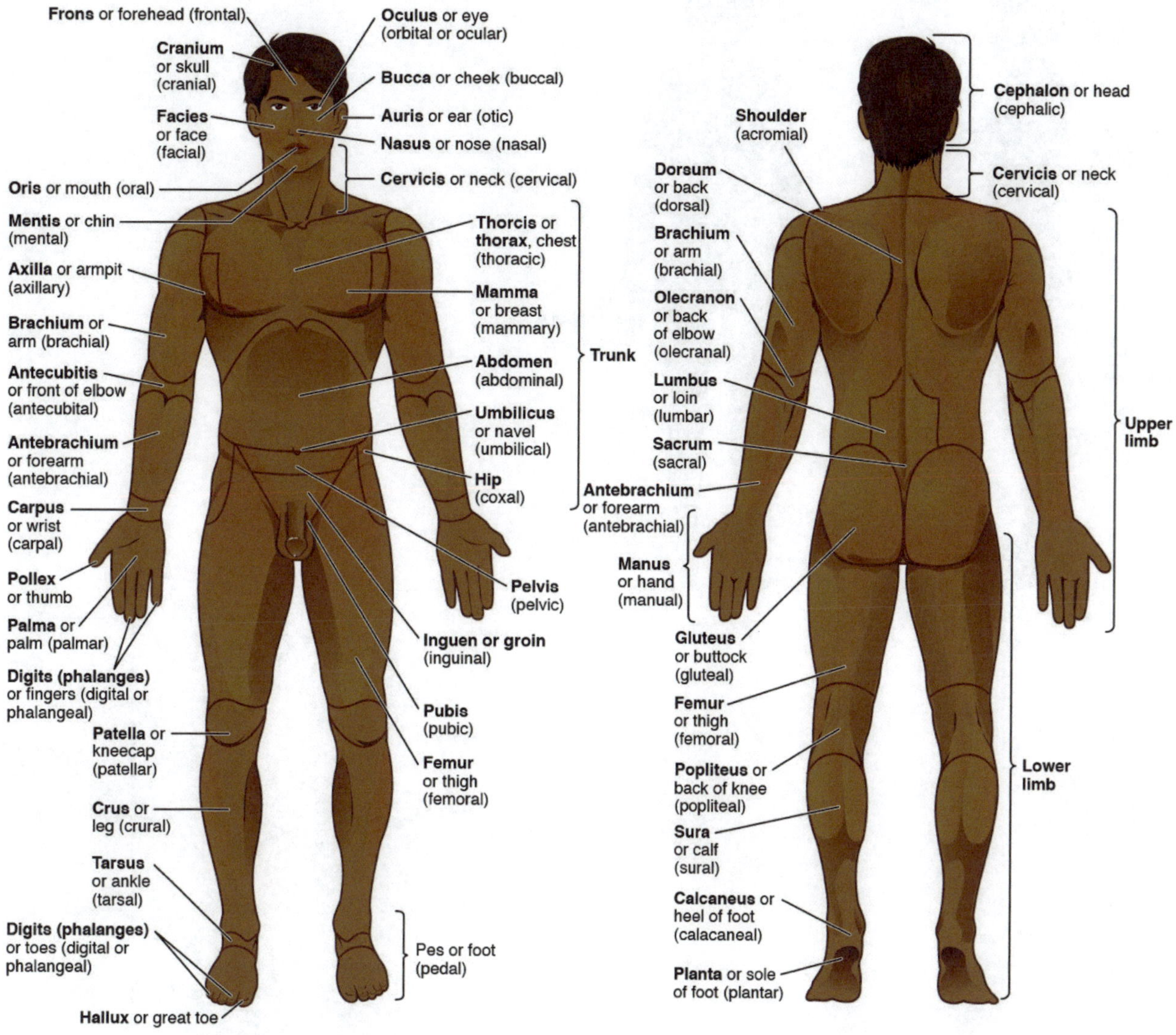


(a) Anterior view (b) Posterior view

[5]

---

[5] "Regional Terms" by OpenStax is licensed under CC BY-SA 4.0 / A derivative from the original work

1

We also have a variety of terms to describe the position in space of structures. These terms we call **anatomical directions**. Since humans are bipedal, our anatomical directions are different than most other mammals. Occasionally we must also split the human body into various planes, especially if we wish to look at internal structures. These are called **body planes**.

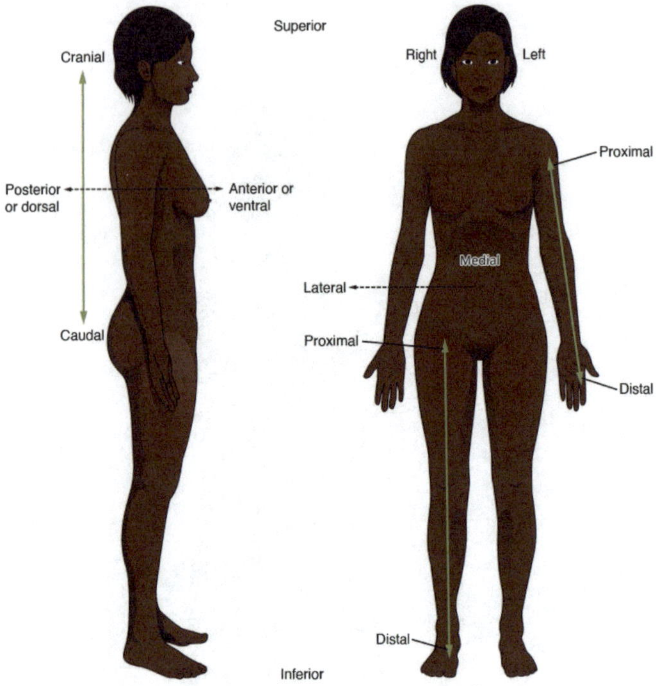


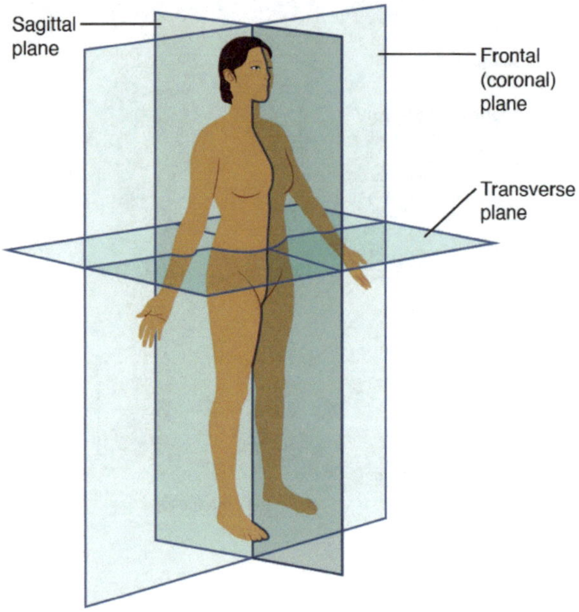


[6] "Anatomical Directions" and "Body Planes" by OpenStax is licensed under CC BY-SA 4.0 / A derivative from the original work

## Review Questions:

1. Many structures are named after the region they are located in. For each structure below, place its corresponding letter on the blank body on the right. You do not have to be exact, but it should be in the correct region.

    a. Biceps **brachii** muscle

    b. Rectus **femoris** muscle

    c. **Inguinal** ligament

    d. **Mental** foramen

    e. Patella bone

    f. Adductor **pollicis** muscle

    g. Axillary nerve

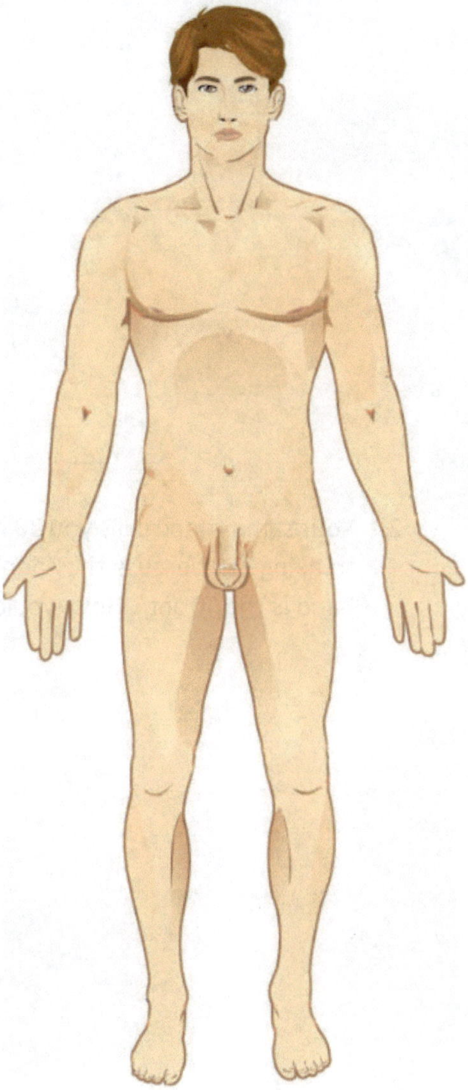

2. For each question use the correct anatomical direction to describe it.

    a. The heart is _____________ to the belly button

    b. The big toe is _____________ to the pinky toe

    c. The ears are _____________ to the nose

    d. The spine is overall a _____________ structure

    e. The breast, belly button, and chin are overall _____________ structures.

    f. Your femur is _____________ compared to the hallux on the same limb

3. Identify the correct plane based on the descriptions below:

    a. Cutting the body into unequal left and right halves would be a _____________ plane.

    b. Cutting the body into a top and bottom half would be _____________ plane.

    c. Cutting the body into a front and back half would be a _____________ plane.

## Application Questions:

[8] "Anatomical Position" by OpenStax is licensed under CC BY-SA 4.0 / A derivative from the original work

1

1. Your friend says that the belly button is superior to the spine, rather than anterior to it. Their logic is that if you are lying down, the heart is above the spine making it a superior structure. What crucial fact about anatomical positions are they missing? (hint: pay close attention to how all the models and images in this section have their bodies placed).

2. Your same friend calls you 30 minutes before your first Anatomy exam. They say they are freaking out because they don't know if the palm of the hand is the "front" or if the back of the hand is the "front". How would you answer their question?

## Activity 5: Organ systems, body cavities and rat dissection

In order to learn about the various organ systems we studied today, you will perform a quick dissection on a rat. However, you will not directly learn rat anatomy. Instead, you will use **comparative anatomy** to identify organs. Comparative anatomy studies how the various structures of different organisms relate to each other. In the case of mammals, essentially all of our organs are in the same locations. If you know where they are in humans, you can likely identify them on a rat. Your goal is to study organs found in the abdominal and thoracic cavities.

First off, grab the following materials:

- Gloves and googles
- Dissecting tray with rubber mat
- Scalpel with blade
- Scissors
- Pointers and tweezers
- Pins

Lay your rat in the **supine** position (on its back). If it does not stay, insert pins through the lateral parts of its body and into the rubber mat. Try to only grab skin with the pins to avoid damaging organs.

You will first be using a scalpel to make cuts through the skin of the rat. Scalpel blades are thin and flexible, but extremely sharp. They are to only be used on soft tissue. You cannot cut through tough cartilage and bone with a scalpel! The blade can break and cut you. Also make sure to always cut away from your hands, the blade can slip.

Using your scalpel blade, make the cuts (dotted lines) below. Your goal is to expose organs found in the abdominal and thoracic cavities. Cut slowly to make sure you do not puncture any organs lying underneath. Once you reach the bones of the ribcage transition to scissors which are capable of cutting through the bone. **DO NOT THROW AWAY RAT UNTIL INSTRUCTED TO DO SO BY THE LAB MANUAL.**

## Try these questions:

Using the labeled human diagram below to help, label the lettered structures on the rat. If necessary, look up a diagram of human body organs to help.

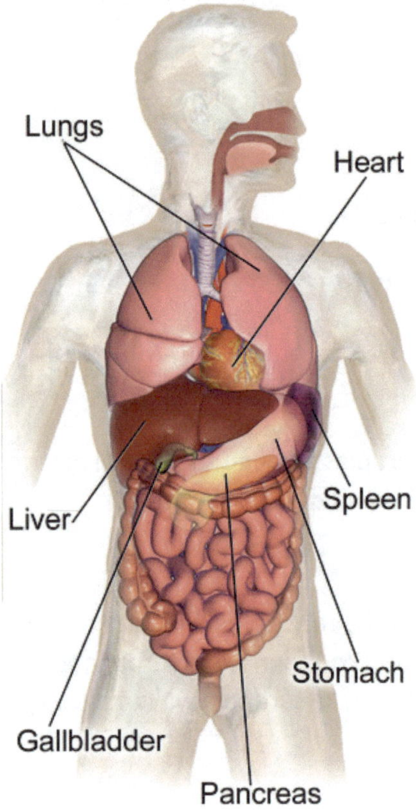


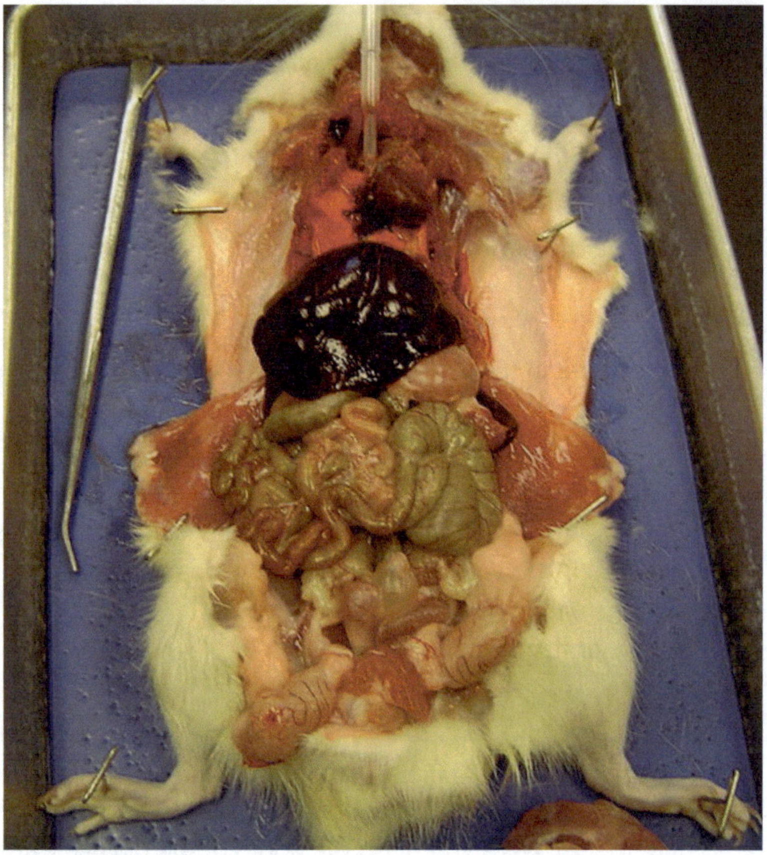

[9] "Cut up rat" by Allen Lew from Berkeley, USA is licensed under CC BY 2.0

[10] "Abdominal Organs Anatomy" by BruceBlaus is licensed under CC BY-SA 4.0.

## Review Questions:

1. Observe your rat, what organs make up the thoracic cavity? Is this the same or different from humans?

2. Observe your rat, what organs make up the abdominal cavity? Is this the same or different from humans?

3. What muscle separates the thoracic and abdominal cavities?

4. Closely observe the heart of your rat. Using a pointer or tweezers, try to find a thin layer of tissue surrounding it. Some may be attached to the rib cage wall. What do we call this thin layer of tissue? What is the general term for this type of tissue? What is its function?

## Application Questions:

1. Pay close attention to the layout of abdominal organs in your rat. Which structures are more dorsal, which structures are ventral? Fill out the table below. <u>You may need to remove organs such as the intestines to see what is dorsal</u>. There is also no distinct cutoff between what is considered ventral and what is considered dorsal, but try your best to abstract under these categories.

| **Ventral Structures** | **Dorsal Structures** |
|---|---|
| | |

2. Imagine you had a second rat, what are some differences you might expect between the rat in the virtual dissection and your second rat? List at least 5. How does this relate to the saying "every body is different".

3. On Canvas there will be a blank rat dissection image. Download this and add it to powerpoint or similar. Add arrows and labels to identify any organs that you see. Upload this powerpoint along with the rest of this lab packet.

# Lab 2

# Cells and the Microscope

## Learning Objectives: Students will be able to…

- Use a microscope
- Create wet-mount slides and smears
- Identify major structures of the human cell
- Estimate the size of cells and other objects viewed under the microscope



## Activity 1: Using a Microscope

The cell is the fundamental unit of life. A single cell is capable of exhibiting all characteristics of life that we are familiar with (e.g., metabolism, growth, reproduction, etc.). For you, however, your cells cooperate together to form tissues, organs, and organ systems. It is than the organisms (i.e., us) that exhibits the characteristics of life. This is because you are a **multicellular organism**. You are made up of multiple cells that work together and can form more complex structures. These larger aggregates are often viewable with the naked eye, but individual cells require a **microscope** to visualize. A microscope is capable of magnifying an object many hundreds of times. This will allow us to see structures of the cell. It will allow us to zoom in on tissues to see much more detail than what is capable with the naked eye.

For this lab's activities you will make use of the "light microscope". It has a range of about 1000μm to 1 μm (**micrometers**). This can also be expressed as 1mm to 0.001mm (**millimeters**). This allows for us to see some of the smallest bacteria up to sections of large tissues such as skin. This is accomplished by passing light through a thin slice or structures (e.g., individual cells). This light is passed through a convex lenses which magnifies the image for our eyes. There are also **electron microscopes** which are capable of magnifying the tiniest structures, and even molcules. As the name suggests, electrons, as opposed to photons ("light") are used to resolve the image.

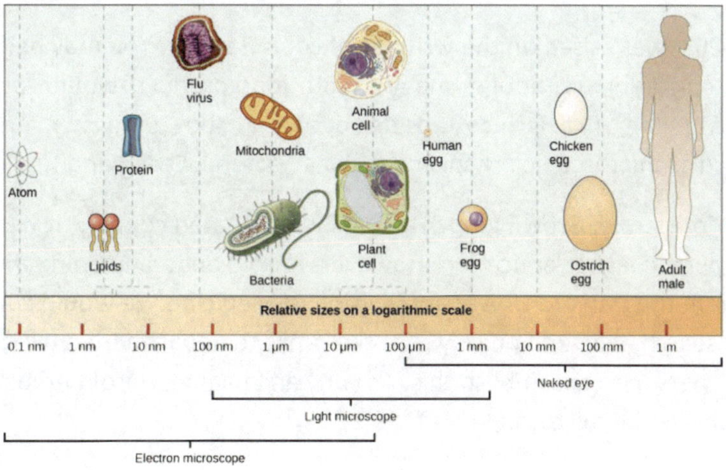

One major issue with viewing cells is that most human cells are colorless. If we want to see the "tiny organs" (**organelles**) that makeup the cell, we need to **stain** the cells. There are many types of stains, but two major categories are **acidic stains**, and **basic stains**. Acidic substances are neutralized by basic subtances, and vice verse. In that regard, acidic stains, stain basic substances such as many proteins of the cell. Basic stains will stain acidic substances such as Deoxyribonucleic Acid (DNA). You will use the common basic stain methylene blue to visualize the **nucleus** of the cell. The nucleus is where DNA, your genetic material, is stored. Here is a selection of other human organelles and structures, and their functions:

Table 1. Common Organelles and Structures

| Organelle/Structure | Function |
|---|---|
| **Cell/Plasma Membrane** | Surrounds the cell and acts as "skin". Allows only certain molecules to enter (**selective permeability**) |
| **Nucleolus** | Site where ribosome (protein creating molecules) are formed |
| **Mitochondria** | Generates energy for the cell |
| **Golgi Apparatus** | Sorts and modifies proteins |
| **Endoplasmic Reticulum** | Translates nucleotides into amino acids |
| **Vacuole** | A "sac"-like structure that can be used to store molecules such as fats |
| **Nucleus** | Storage site for genetic information (i.e., DNA) |
| **Cilia** | Used for propulsion of substances over the surface of a cell; the cell itself is stationary |
| **Microvilli** | Increase surface area to aid in absorption |
| **Lysosome** | Vesicle that contains digestive enzymes to break down food particles |
| **Extracellular Matrix** | A network of fibers found outside of the cell that link the inside and outside |
| **Cytoplasm** | The fluid inner contents of the cell |

In this first activity, you will use the microscope using **prepared slides**. This slides have already been made for (prepared), so all you need to do is place them on the stage of the microscope and focus.

## Every-body is Different: Being Black in the Academy

Harold Amos (1918-2003) was raised on the writings of Louis Pasteur. You may be familiar with Pasteur for his process of pasteurization. Without it, many of your food products would spoil from growth of bacteria. It is no surprise then, that Harold Amos decided to pursue studies in **microbiology**. This subfield of biology studies microscopic organisms such as bacteria, yeast, and fungi.

This was no easy task for Harold. Growing up in a heavily segregated country, it was almost impossible for a Black person to pursue higher education. However, Harold ended up being the first Black person to receive a doctorate from the Harvard University Division of Medicine. He would continue to pursue microbiology as his research field, and quickly rose the ranks to department chair at Harvard Medical School- The first Black person to accomplish this. As you can imagine, Harold Amos's success depended heavily on his ability to use the microscope

If you follow these steps, you should be able to resolve any sample. Particularly important steps are bolded. A microscope diagram is on the right to help you.

1. Plug the microscope in and turn the microscope on. Adjust the brightness as necessary.
2. **Adjust the condenser aperture lever so that the max amount of light is passing through.**
3. Place the slide on the stage. Lower the stage all the way down using the coarse focus if it isn't already.
4. Set the objective lens to 4x
5. Adjust the X/Y position of the stage so that the sample is being illuminated by the light source.
6. With the objective on 4x (the lowest power), adjust the coarse focus until the sample comes into focus. **Do not move on until you have focused at 4x!** There is no risk of breaking your slide by ramming it into the objective lens, so you can use your coarse focus freely.
7. Move up to the next objective lens. **Refocus using only the fine focus.** Repeat at higher magnifications if needed. The highest magnification, 100x, is an oil immersion lens. A droplet of oil is placed on the slide and the lens is immersed in the oil droplet. **For anatomy, you won't likely have to use the oil immersion lens.**

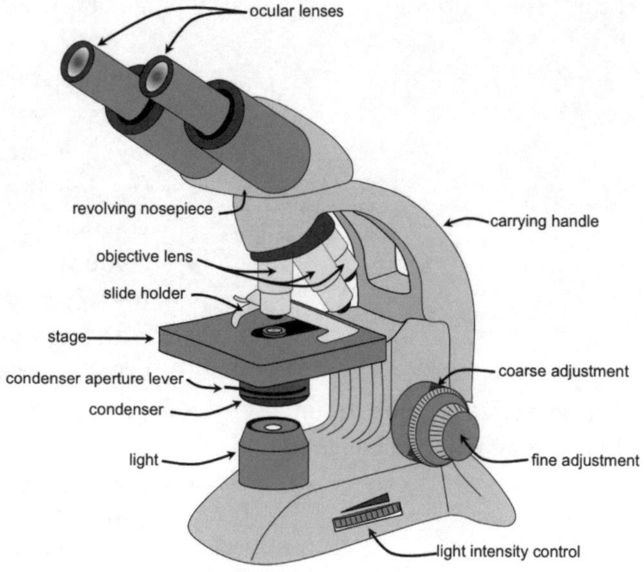


## General Care and Cleanup of the Microscope:

Below are a variety of guidelines that need to be followed when using the microscope.

- Always handle with care, use two hands when carrying
- Never force anything. If it feels stuck let me know.
- If you see lint on the lens or if you have used immersion oil, **pat (not wipe) the lens with lens paper and cleaning liquid only**. Do not let the cleaning liquid air dry.

For cleanup after using the microscope, do the following:

1. Lower the stage
2. Make sure the microscope is turned off and the brightness is lowered completely
3. Return slides to their appropriate position in the slidebox
4. Wrap the cord loosely around the base of the microscope return
5. Return the microscope to its correct position in the cupboard

## Troubleshooting:

Below are common issues with using the microscope. Consult these first, and if your problem still can't be fixed, ask me for help.

Table 2. Microscope Troubleshooting Tips

| Issue | Solution |
|---|---|
| **I can't see anything!** | Assuming your microscope has power and lights up, try these....<br>• **Did you start at 4x?** return to that magnification and try to focus on your sample before moving up to the next level.<br>• If you are at 4x, **try raising your stage up**. The majority of the time the stage is too low. Make sure to go slowly, so you don't miss the point at which your sample is in focus.<br>• If you've tried raising and lowering your stage, **try adjusting the X/Y position of your slide.** You may not be over a spot with a sample. Samples are covered with a cover slip, and usually died pink, purple or blue. |
| **My objective is going to run in the slide when I change it!** | • **Did you start at 4x?** return to that magnification and try to focus on your sample before moving up to the next level.<br>• **Did you adjust the coarse focus after focusing at 4x.** Once you've focused at 4x, you should never have to adjust the coarse focus again. The fine focus only should be used. |
| **Everything is dark, or hazy, or faint!** | • **Turn up the brightness**<br>• **Adjust the condenser aperture lever.** This will change the contrast of the image.<br>• **Try another slide.** Some slides have faint staining that makes them hard to see regardless of setting. |

## Activity 1 Procedure:

1. Reset your miroscope from any work you've done earlier
   a. Change to the lowest power
   b. Lower the stage
   c. Remove and slides on the stage and return to their appropriate box
2. Grab an "e" slide from the front of the classroom
3. Place it on the microscope stage so that the "e" is facing you (as if you were trying to read)
4. Focus using the lowest power, 4x
5. Draw the "e" <u>exactly as you see it,</u> to the right



6. We have plenty of other prepared slides at the front of the classroom. Skim through these and find a couple that interest you. View these on your microscope and draw them below. Make sure to list the title of the slide and magnification that you drew it at.

## Review Questions:

1. Name each part of the microscope below

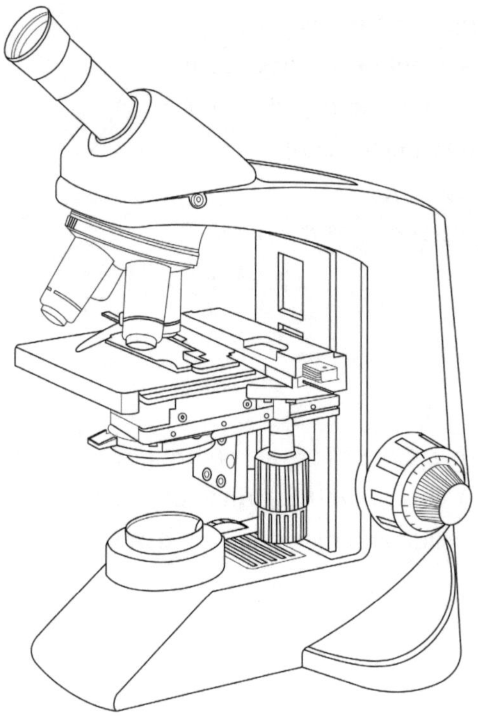

2. Each statement below has one wrong word in it. Cross out the incorrect word and write in the correct one.
    a. The microscope power cord should be wound tightly.
    b. The stage should be raised before storing the microscope.
    c. Turn the brightness up on the microscope before storing it.
    d. The objective lens should be at the highest magnification once finished.

## Application Questions:

1. If you followed steps 1-5 of the procedure correctly, you likely noticed something weird about the orientation of the letter "e" when viewing it under the microscope. What happened to the letter "e"? What does that tell you about any sample you view under the microscope in the future?

## Activity 2: The "Compound" Microscope

Even though a microscope has objective lenses that magnify the image, the ocular lenses (the ones you look through) also have a magnification. This magnification is a constant 10x. Sometimes you may hear a light microscope referred to as a **compound microscope** (as in the blank image on the previous page). This is because the objective and ocular lens magnifications are compounded (multiplied). Calculate the total magnification for each objective lens using the table below:

2

| Objective Lens | | Ocular Lens | Total Magnification |
|---|---|---|---|
| 4x | multiplied by... | 10x | 40x |
| 10x | multiplied by... | 10x | |
| 40 | multiplied by... | 10x | |
| 100x | multiplied by... | 10x | |

## Application Questions:

1. Imagine you had a microscope with 32x objective and 11x ocular lens. What would be the total magnification?

2. Are all light microscopes also compound microscopes? Are all compound microscopes also light microscopes? Explain your answer.

## Activity 3: Cell size

The image on the first page should give you an idea of approximate sizes of cells. However, in this activity, you will develop the ability to estimate cell size for the exact cells you are looking at.

First, place one of the clear plastic rulers on the stage and make sure you are viewing it at 4x magnification.

How many millimeters do you count? Record this here_______________ mm

You have now successfully measured your field of view. It is ________mm (see your answer above) across; You know from one edge to the other edge at 4x is ________mm.

Now, go to your 10x. At this magnification, you likely won't be able to see the markings on the ruler clearly because you are in between them. You must then use math to calculate your field of view.

10x is 2.5 times more zoomed-in than 4x. You are seeing 2.5 times less than 4x. This means your field of view is _______________ mm.

Repeat the same math above for 40x and 100x

40x has a field of view of _______________ mm

100x has a field of view of _______________ mm

With the above in mind, we can estimate the sizes of cells using the known field of view. Imagine we have a drawing such as below. We can approximate that 2.5 clown shoes across will fit in our field of view. That means the length of 1 clown shoe is your field of view divided by 2.5. You can repeat the same steps for height where ~5 clowns shoes stacked will fit our field of view. Again the field of view divided by 5 will give the height of a single clown shoe. This is not an exact way to calculate cell size, but it will give a good estimate.

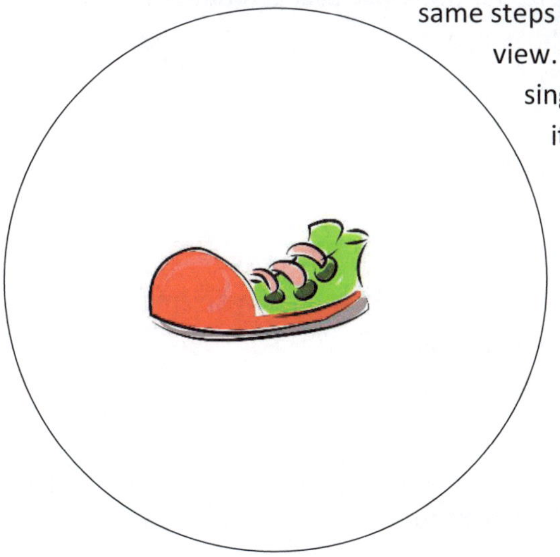

## Activity 4: Making wet mount slides

While some biological samples are sold as prepared slides, the vast majority are not. You will have to make your own slides. For ones that will have the sample immersed into water, this is called a wet-mount slide. Follow the directions below, and/or or use this instructional video (https://www.youtube.com/watch?app=desktop&v=RfaSyfDBQzU)

2

Note that different instructions use different tools for laying the coverslip on the slide. Any of these are fine. Your fingers can be used as well as long as you handle the edge of the coverslip. The same goes for the slide itself, only handle the edges. We don't want any smudges to obstruct your view.

### Activity 4 Procedure:

1. Make a wet mount slide of an elodea leaf
    a. Grab a fresh slide and lay it on a clean spot on the benchtop
    b. Add a drop of water
    c. Using tweezers, grab one of the elodea leaves and place it in the water
    d. Lay a coverslip on the edge of the water and slowly lower it
    e. Use steps from activity 1 to view the sample, and draw it on the next page. Label the name of the sample and total magnification next to the circle you use. Estimate the height and length of the cell and record this on the image.
2. Make a wet mount slide of an onion slice
    a. Repeat steps 1a and 1b
    b. Cut a **very** thin slice of onion
    c. **Add a drop of iodine solution**
    d. Repeat steps 1d and 1e
3. Make a wet mount slide of pond water with protoslo additive to slow down the organisms
    a. Grab a fresh slide and lay it on a clean spot on the benchtop
    b. Add a drop of pond water and one drop of protoslo solution
    c. Follow steps 1d and 1e
4. Make a wet mount slide of your cheek cells
    a. Grab a fresh slide and lay it on a clean spot on the benchtop
    b. Add a drop of water

c. Scrape the inside of your cheek (i.e. your mouth) gently with the toothpick and dip it in the water. Swirl the toothpick in the water.
d. Dispose of your toothpick in the specified container
e. Add a drop of methylene blue
f. Repeat steps 1d and 1e above.

5. Once you are done drawing, place the slide and sample in the “used slides” container

2

Intentionally left blank

# Lab 3

# Epithelial Tissue

## Learning Objectives: Students will be able to...

- Explain what a "tissue" is
- List all types of epithelial tissue, and where they can be found
- Compare and contrast simple vs. stratified epithelium
- Identify various types of epithelial tissue when shown a slide or image

## Activity 1: Epithelial tissue

**Epithelial tissue** is perhaps a tissue you are very familiar with. Every time you are looking at your skin, you are looking at layers of dead epithelial tissue. Epithelial tissues are known for lining surfaces, but these surfaces can be both internal and external. For example, the inside of your blood vessels is lined with epithelial tissue. This is confusingly called **endothelium**. Similarly, your abdominal cavity is lined with a serous membrane made of epithelial tissue. Beyond lining surfaces, epithelial tissue serves a variety of functions in the human body, such as **exocrine glands** and absorption. The former excretes substances to the outside environment (e.g., sweat, milk, etc.).

It is perhaps best not to think of epithelial in terms of where it is found, but instead the characteristic structure that comprises it. Epithelial tissues are organized into two major groups: **simple and stratified epithelial tissues**. Simple epithelial tissues are a single layer of cells bound to a **basement membrane**. Think of this as an anchoring point for the issue. Binding the cells laterally together are **tight junctions** that prevent molecules from slipping in between the cells. Stratified tissues are similar in that they connect to a basement membrane, but now they are multiple layers thick.

In addition to being simple or stratified, epithelial tissues also have a shape. Flat cells are called **squamous** – They look like a square when viewed from above. **Columnar** cells are taller than they are wide – they look like columns. Finally, **cuboidal** cells have equal height and width – they are cubed shape. If we combine the number of layers with the shape, we get the names of various epithelial tissues (e.g., simple squamous, stratified cuboidal, etc.). There are also 2 more shapes that will be discussed later: transitional and pseudostratified. These are odd exceptions to the 3 shapes above.

What is a **tissue** though? Tissues are a group of cells working together to perform a similar function. Your skin is an example of multiple cells (keratinocytes, specifically) working together to waterproof and protect your body. Combinations of the 4 major tissues epithelial, connect, neural, and muscular, then go on to form organs. It is from this microscopic foundation that all the structures of your body are formed. We call this study of the microscopic structures of the human body **microanatomy**, or **histology**.

## Simple Squamous (1)

Despite the presence of tight junctions, the flat shape of simple squamous tissue allows for molecules such as oxygen to quickly diffuse across the plasma membrane. Remember, the plasma membrane is selectively permeable, only certain molecules can pass through without the help of membrane proteins.

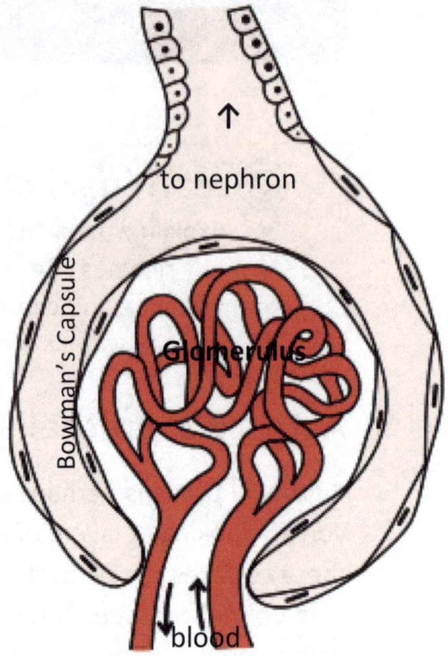


One good example of simple squamous epithelium is in the diffusion of extracellular fluid across the membrane into the nephrons that make up the kidney. You can think of the kidney as being a large filter for the human body. The liquid excreted by the kidneys (urine) is then an excellent reflection of what is going on internally. Molecules such as hormones, drugs, and nutrients found in the blood are filtered across the simple squamous tissue that makes up Bowman's capsule. Extracellular fluid is provided by capillaries known as glomeruli which provide blood.

**When looking for Bowman's capsule you want to find a sphere (the capillaries or glomeruli) surrounded by a gap and then a thin layer of cells (simple squamous tissue or Bowman's capsule).** The glomeruli can be easily identified by the many dark purple nuclei on it (cells called podocytes cover the capillaries).

Be careful though! As you travel further down the nephron the tissue turns into simple cuboidal cells (see the image to the right). These tubes bend and turn, so you may see a circle of simple cuboidal tissue. Notice there isn't a glomerulus in the center of these cuboidal cells.

---

[1] Public Domain

**Draw what you see:**

Make sure to label the following: glomerulus, Bowman's capsule, simple squamous cells, kidney tubule, simple cuboidal cells

Total Magnification of Drawing:



## Simple Cuboidal Cells (2)

You can imagine a cuboidal cell has a greater volume than flat, squamous cell. Their nucleus then has more space to be spherical. Simple squamous tissue though is so flat that the nucleus must take that shape as well; they get squished!

One such example is the thyroid gland which is made up of circles of simple cuboidal cells (much like the kidney tubules you saw above). These circles actually make up a 3d sphere called a thyroid follicle. Another function of simple cuboidal cells is in the formation of various "tubes" or "containers" of the human body (see above for its use in the kidney). By forming a circle or sphere they can develop these functions. For the thyroid gland, the colloid contained by these simple cuboidal cells contains thyroglobulin, the precursor molecule to thyroid hormones.

---

[2] Arrangoiz et al. (2018). Used by Permission. Creative

Commons Attribution International License

These follicles should be easy to spot on this slide, but don't confuse them with stratified cuboidal cells! Many follicles butt up against each other giving the illusion of 2 layers of cuboidal cells. If you look closely enough, you will notice that these 2 rows of cuboidal cells are actually separated by a dark staining basement membrane. There are also triangles at junctions of various follicles which also uncharacteristic of a stratified tissue.

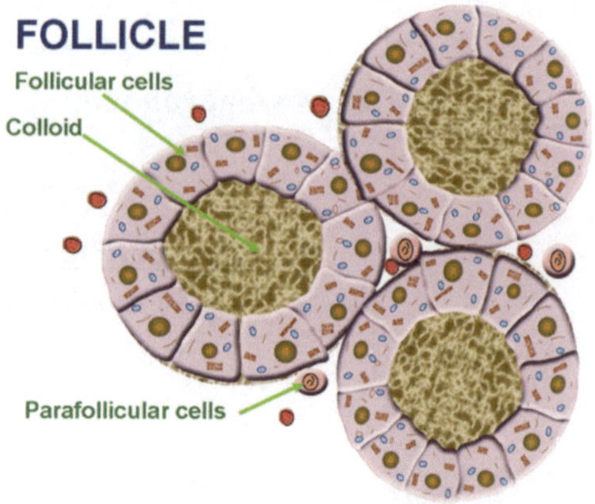


**Draw what you see:**

<u>Make sure to label the following:</u> thyroid follicle, simple cuboidal cells (follicular cells), colloid



Total Magnification of Drawing:

[3] Weupe et al. Used by Permission. Creative Commons License

## Simple Columnar (3)

These cells are very similar in function to simple cuboidal. One major difference is location. Unlike simple cuboidal cells, simple columnar cells are found particularly in the intestine.

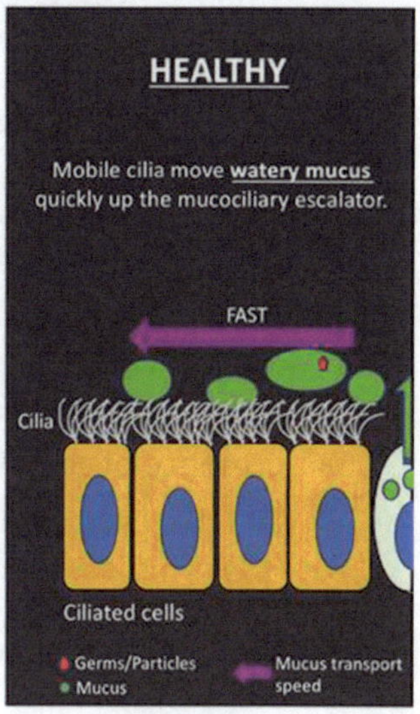


One special feature of simple columnar epithelium is that they may have apical modifications (e.g., usually microvilli for simple columnar and cilia for pseudostratified columnar). For the case of cilia, they aid in movement of particles by beating like the oar of a boat. For microvilli, they increase surface area which aids in absorption such as in the small intestine. The height of these cells, in part, allows for this specialization.

**Draw what you see:**

Make sure to label the following: simple columnar epithelium

Total Magnification of Drawing:

## Pseudostratified Columnar (4)

The prefix pseduo- means false. So pseudostratified columnar epithelium, is a false stratified tissue. Each cell has varying height, so when viewed in cross section it appears as though the tissue is stratified (made up of multiple layers). However, each cell in this tissue is touching the basement membrane, making it, in fact, a simple epithelial tissue. Compared to simple columnar tissue the nuclei of pseudostratified columnar tissue has nuclei of varying heights. The nuclei of simple columnar tissue are generally of equal height.

Just like simple columnar tissue, pseudostratified columnar tissue has apical modifications. This time in the form of cilia. The most common example of this in the trachea ("wind pipe") leading to your lungs. The cilia of the tissue beat which causes mucus and trapped particles to move up and out of the trachea. Notice that underneath the pseudostratified cells, we have connective tissue (named the lamina propria here). Do not confuse this with what you trying to look for.

Another important aspect, especially of tracheal pseudostratified columnar epithelium is the presence of goblet cells. These simple columnar cells contain mucous which is released into the tracheal lumen. These cells are easily spotted since the vesicles containing the mucous do not stain well with H&E. This leaves a large, light pink circles within the goblet cell.

**Draw what you see:**

Make sure to label the following: ciliated, pseudostratified columnar epithelium, cilia, goblet cells, lamina propria (connective tissue)



Total Magnification of Drawing:

## Stratified Squamous (5)

Stratified tissues are multi-layer tissues where only the deepest layer is in contact with the basement membrane. In naming the tissue, it is important to pay attention to the most superficial cells. As you will see with stratified squamous tissue, the depp cells are cuboidal in shape, but the superficial cells are squamous (flat).

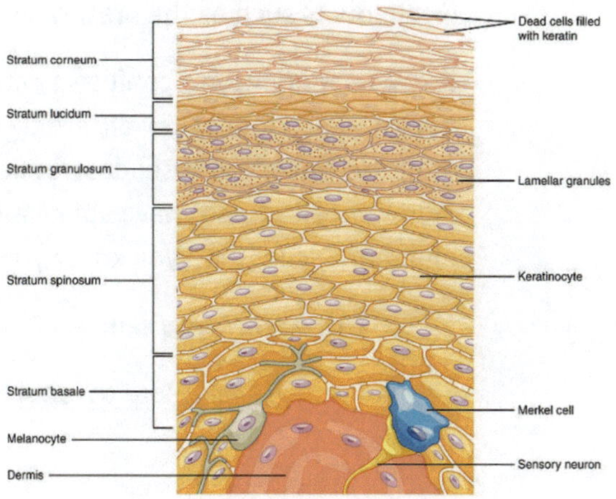


For example, view the skin diagram shown on the right. The superficial layers are composed of extremely flat cells and make up the layer of the epidermis called stratum corneum. As we move down the picture to more deep cells you will see they start take more cuboid shapes. In naming the tissue it is again important to pay attention to the most superficial layer of cells which in this case are flat or squamous in shape; this is where we get the name for this tissue.

Underneath the deepest layer (stratum basale) is a basement membrane along with connective tissue. As you go even deeper you may even see adipose tissue which makes up the subcutaneous layer.

**Draw what you see:**

Make sure to label the following: stratified squamous tissue, basement membrane, connective tissue

Total Magnification of Drawing:

[4] Access for free at https://openstax.org/books/anatomy-and-physiology/pages/1-introduction. Used by Permission. CC Attribution 4.0 International

## Stratified Columnar (6)

This is a rare tissue type like stratified cuboidal tissue. Usually, it is found in only the largest portions of certain ducts such as the urethra and those of the salivary gland. Like stratified cuboidal tissue, it is rare.

There is a tendency to confuse stratified columnar with pseudostratified columnar, a simple tissue. There are a couple of key differences. One, stratified columnar is usually only two layers of cells; if we to count the "layers" of pseudostratified columnar cells we would end up with many more. Two, stratified columnar cells do not have apical modifications. If you see cilia on something that looks like stratified columnar tissue, then you know you are actually looking at pseudostratified columnar epithelium.

**Draw what you see:**

Make sure to label the following: stratified columnar tissue, basement membrane



Total Magnification of Drawing:

## Stratified Cuboidal (7)

Unlike stratified squamous tissue which is abundant (it makes up part of your largest organ, skin), stratified cuboidal tissue is comparatively rare. It is found in the anorectal junction, and the ducts of various glands such as sweat and mammary glands. In most cases stratified cuboidal tissue is only two layers thick.

**Draw what you see:**

Make sure to label the following: stratified cuboidal tissue, basement membrane

Total Magnification of Drawing:

## Transitional (8)

Transitional epithelium is similar to stratified squamous epithelium upon surface evaluation. However, one major difference is that transitional epithelium is able to stretch or "transition". When relaxed, the superficial cells have a rounded shape (see picture), when distended these cells flatten.

The most famous example of this tissue in the bladder which utilizes the stretching feature of transitional epithelium. It is also found in other regions of the urinary system, but nowhere else in the body.

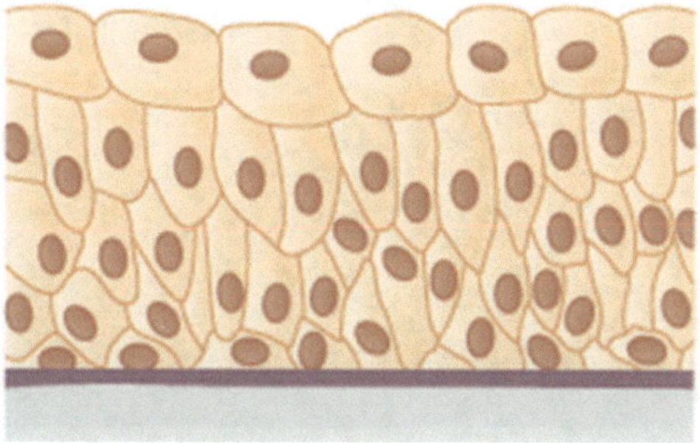

When distended, transitional epithelium can often be confused with stratified squamous epithelium. There are [5]few major differences such as transitional epithelium never being

[5] Jmarchin. Used by Permission. CC Attribution Share-Alike 3.0 Unported.

keratinized. Also, transitional epithelium generally has fewer layers and cells than stratified squamous epithelium.

Make sure to label the following: transitional epithelium

Total Magnification of Drawing:

## Review Questions:

1. Fill out the following concept map that reviews epithelial tissue.

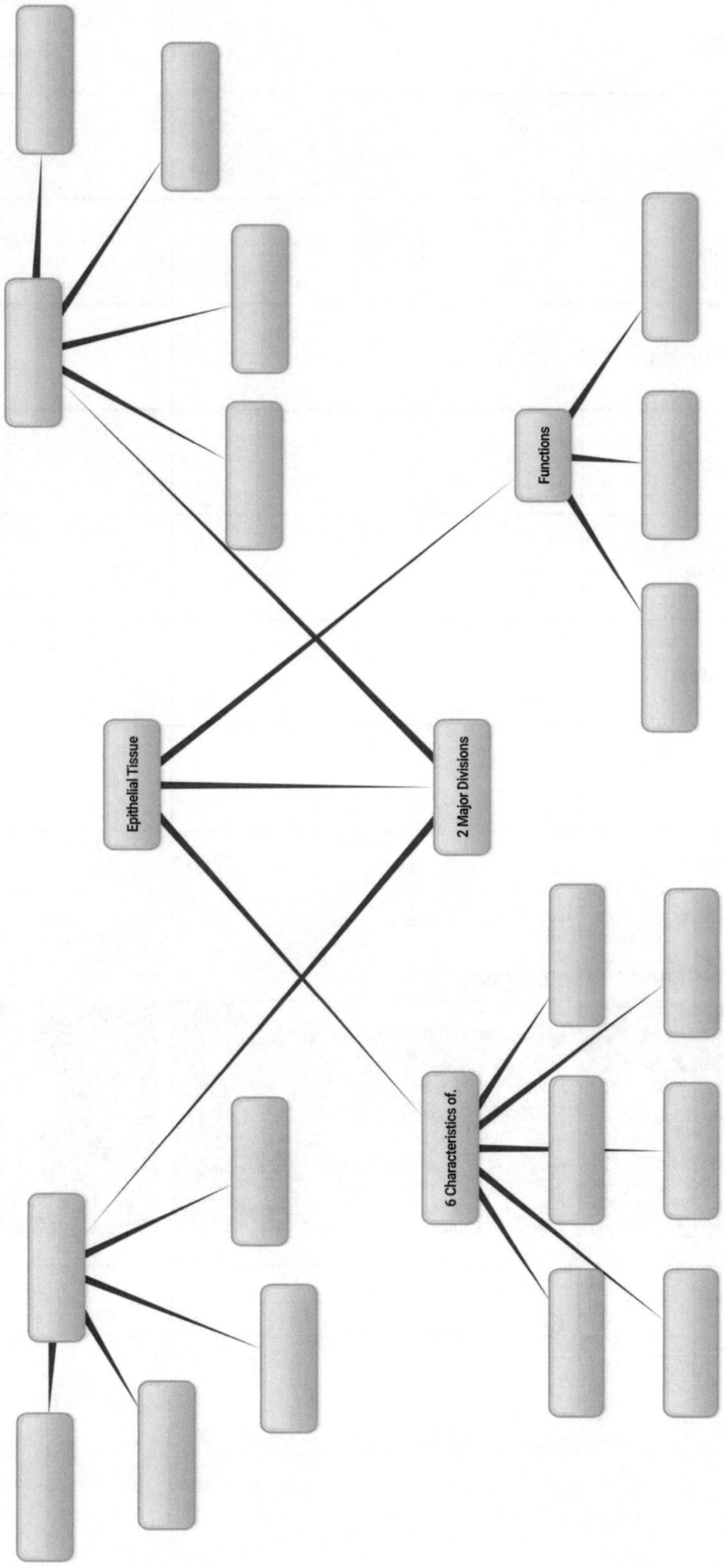


3

2. Fill out this table with examples of where each epithelial tissue is found, and its general functions.

| Tissue | Locations in Body | Function |
|---|---|---|
| Simple Squamous | | |
| Simple Cuboidal | | |
| Simple Columnar | | |
| Pseudostratified Columnar | | |
| Stratified Squamous | | |
| Stratified Cuboidal | | |
| Stratified Columnar | | |
| Transitional | | |

## Application Questions:

1. Identify each epithelial tissue below:

| | | |
|---|---|---|
| 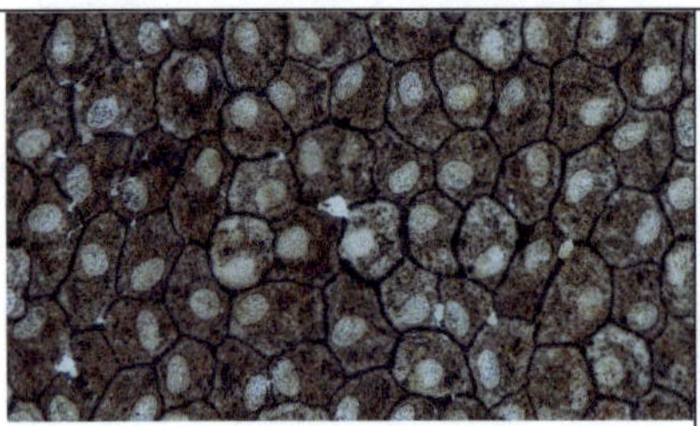 (Top down view) | 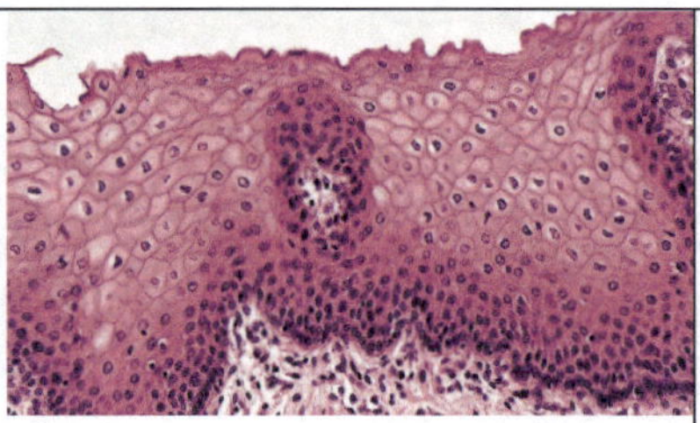 | 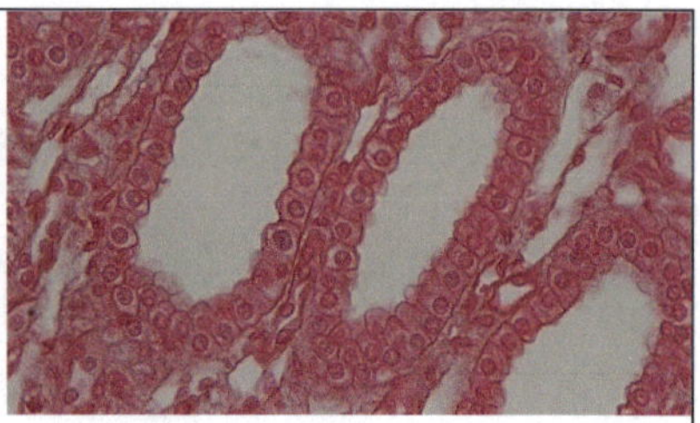 |
| | | |

# Lab 4

# Connective Tissue

## Learning Objectives: Students will be able to…

- List all types of connective tissue, and where they can be found
- Identify various types of connective tissue when shown a slide or image
- Identify any italicized structures listed in this lab
- List some common cells found in connective tissue and their function

## Activity 1: Connective tissue

**Connective tissue** is considerably more complex than epithelial tissue in terms of structure and function. The major defining characteristic of connective tissue is its complex **extracellular matrix** which is composed of a ground substance (e.g., proteoglycans, glycoproteins, and glycosaminoglycans) and a protein fiber (collagen, reticulin, and/or elastin). It is usually the extracellular matrix that determines how the connective tissue type is described and named. This large extracellular matrix means that connective tissue generally has fewer cells than epithelial tissue. There is simply not enough space to have the same density of cells that epithelial tissue has.

One of the benefits of more space is that there can be **wandering cells**. As the name suggests, these cells can move. They are not bound by tight junctions, or have such little space that they cannot migrate. One excellent example of this is the **macrophage**. Macrophage means "big eater" in Latin. In response to inflammatory molecules such as those released due to a cut in your skin, these macrophages will migrate ("wander") towards the wound.

Connective tissue can roughly be divided into three major categories: proper, liquid and supportive. Proper connective tissue includes types such as dense regular, and areolar, liquid connective tissue is blood and lymph, and supportive connective tissue is composed of various types of cartilage and bone. The functions of these tissues are diverse; the word "connective" cannot begin to describe the numerous and specific functions they perform.



## Clinical Application: Collagen Injections

Collagen is certainly one of the fibers of connective tissue that you have heard of. It is often advertised as helping skin to look more beautiful. When injected into the dermis of the skin it can help to reduce wrinkles by plumping the skin. In the body, cells called fibroblasts are constantly building new extracellular matrix (i.e., collagen). However, in aging skin, these cells stop functioning and/or die off. This reduces the amount of collagen in your skin and is one factor that contributes to wrinkles. Connective tissue is essential for the structural integrity of our bodies, and we can see that even with our own naked eyes.

## Try this:

1. Review your lecture notes to complete this concept map.

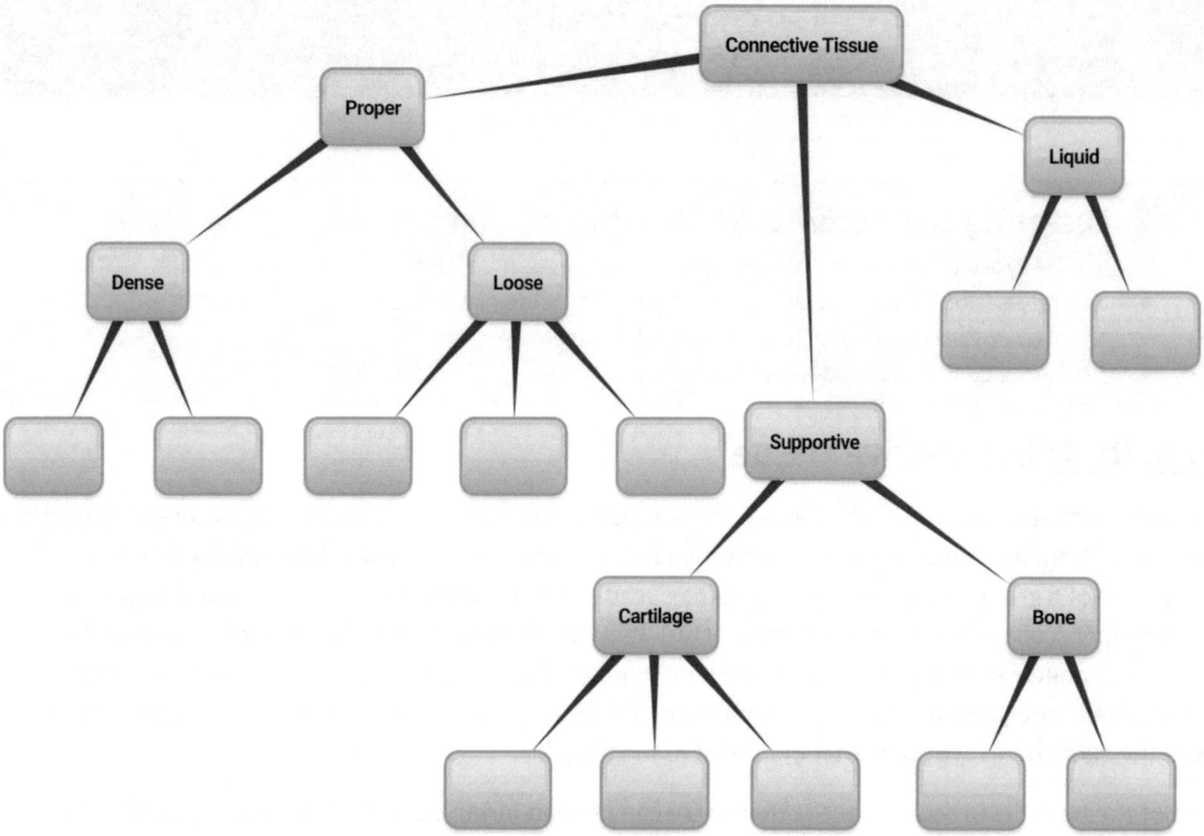

## Areolar (9)

We will first take a look at areolar connective tissue. While not the first type of connective tissue on our concept map above, it is probably the most important for outlining some important features of connective tissue. The most critical is that it contains all three protein fibers: collagen, reticulin, and elastin.

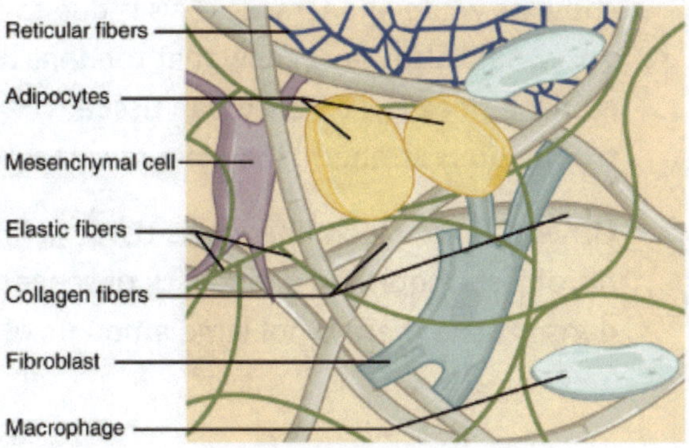


When looking at your slide you may notice that you can only differentiate two fiber types: collagen (the thick fibers) and elastin (the thin fibers). The third fiber, reticulin, does not stain well with H&E which is likely what the slide you are viewing has been stained with. A silver stain is usually used to visualize reticulin fibers (more on this later).

Being a loose connective tissue, there is plenty of room for wandering connective tissue cells to move around within this tissue. In the picture above you can see various cells including macrophages (immune cells), adipocytes (fat cells), and mesenchymal cells (connective tissue stem cells). Note that this image is an example, not a stereotype. You may or may not see these cells or others under the microscope. It is highly dependent on the age, health, and other numerous factors of the tissue donor. At the very least you will see collagen, elastin, and the numerous nuclei of fibroblasts.

**Draw what you see:**

Make sure to label the following: *fibroblasts*, *collagen, elastin*, and any other cells you can identify (pay close attention to their shape!)

Total Magnification of Drawing:



---

[1] "Connective Tissue" by OpenStax is licensed under CC BY 3.0

## Dense Regular (10)

If your slide box contains a particularly old slide, you may see it labeled as "white fibrous connective tissue". Many decades ago dense regular tissue was referred to by this name. This is an excellent description since it is white, and definitely fibrous. If you have ever prepared meat you will notice the silvery tendons attached to it. Tendons are a prime example of dense regular connective tissue, and show its white appearance. This is due to numerous collagen fibers which are colorless (unless stained). You would also likely know that tendons are not something you want attached to the meat you are cooking since it is a very tough tissue. The collagen fibers all run in the same direction which gives it tremendous strength, the main reason why it used to hold muscle to bone in tendons.

Other than collagen fibers, this tissue is fairly simple. You may notice the dark staining nuclei of fibroblasts (squished due to lack of space), but beyond that it is devoid of other cells types. Being a dense tissue there is not large amounts of space for wandering cells.

**Draw what you see:**

Make sure to label the following: *fibroblasts*, *collagen*



Total Magnification of Drawing:

## Dense Irregular (11)

Dense regular tissue gets its name from the *regular* ("neat") organization of collagen fibers. Dense irregular tissue on the hand has an irregular ("disorganized") pattern of collagen fibers. While this tissue is still strong, it is not as strong as dense regular tissue. The benefit is that it has some give to it. Organs such as skin and eyes which need to handle some compression and expansion benefit from the flexibility (and strength) of dense irregular connective tissue. You may notice more spherical nuclei of the fibroblasts due to the availability of space in this tissue.

**Draw what you see:**

<u>Make sure to label the following:</u> *fibroblasts*, *collagen*

Total Magnification of Drawing:



## Adipose Tissue (12)

Returning back to loose, proper tissues we have adipose tissue, or "fat tissue". This tissue type you are all familiar with. White fat (the most common fat in adult humans) is mainly used for energy storage. In turn, the bulk of the cells are vacuoles full of lipid droplets. These lipid

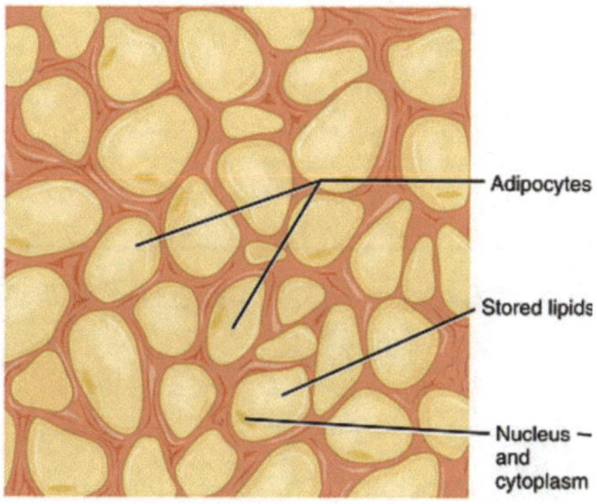


droplets do not stain well with H&E (or really anything else), and are washed way during tissue preparation anyways. This give this tissue its spiderweb-like appearance. The nucleus, mitochondria, and other organelles that stain pink and purple are then squished to the side of the cell. Ocassionally there are fibroblasts found in this tissue producing collagen, but a very small amount of it.

Be careful, you may have viewed a simple squamous slide derived from lung tissue. It will have a similar spider-web like appearance, but each white circle will be surrounded by numerous nuclei. Here it is one nucleus per white circle (lipid vacuole).

Brown adipose tissue which you may have heard of is common in newborns, but not adults. Whereas white adipose tissue for calories, brown adipose tissue is for heat. This is why animals that hibernate tend to have lots of brown fat.

**Draw what you see:**

Make sure to label the following: *adipocyte, lipid vacuole, nuclei, blood vessel*

Total Magnification of Drawing:



## Reticular (13)

As described above, reticular tissue does not stain well with H&E, the most common histological stain and likely what you have seen the most of thus far. In turn, reticular tissue will be an odd duck in terms

[2] "Adipose Tissue" by OpenStax is licensed under CC BY 3.0. Image cropped.

of appearance. It will likely be stained grey, green or black (but NOT always) depending on the stain. Sometimes a counterstain will be used to help visualize the nuclei of cells.

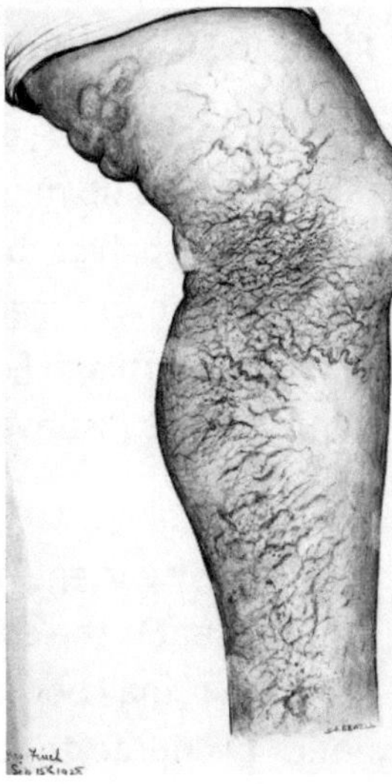

Reticular fibers form a mesh pattern somewhat reminiscent of varicose veins commonly found on the leg. Functionally, reticular tissue provides a “soft skeleton” for certain organs such as the spleen and lymph nodes.

**Draw what you see:**

Make sure to label the following: *reticular fibers, fibroblasts*

Total Magnification of Drawing:

## Hyaline Cartilage (14)

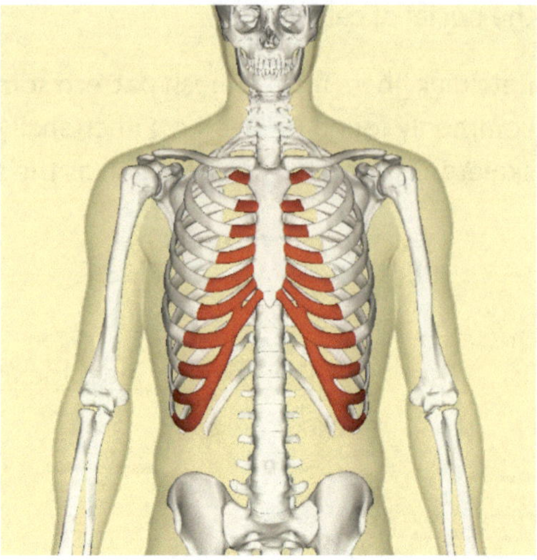

Hyaline cartilage is sometimes called "baby cartilage". It indeed makes up much of our "skeleton" when we are fetuses and babies, but we find it in adults as well. For example, the cartilaginous parts of your rib cage are made up of hyaline cartilage (we call these pieces of "costal cartilage", but really it is just hyaline cartilage).

Hyaline cartilage itself is surrounded by a perichondrium (an area of dense irregular connective tissue). The dominant cell type present is the chondrocyte which is surrounded by a lacunae, a small pocket within the collagen matrix. You may notice chondrocytes clustered together into isogenous groups. These are chondrocytes that have recently divided. As they produce matrix, it will push them away from each other. Like connective tissue proper, collagen is the main fiber found here, however it takes on a diffuse characteristic rather than the prominent fibers found in tissues such as dense regular. This collagen gives hyaline cartilage its strength.[4]

**Draw what you see:**

Make sure to label the following: *perichondrium (dense IR), chondrocytes, lacunae, isogenous group (if present)*



Total Magnification of Drawing:

[4] "Costal cartilages" by Anatomography is licensed under CC BY-SA 2.1 Japan

## Fibrocartilage (15)

Particularly strong, this cartilage makes up your intervertebral disks. For a variety of reasons, these discs may become herniated ("bulge") and compress ("pinch") a nerve causing pain, numbness, tingling, and/or muscle weakness. With their role of helping to support the spine, this type of cartilage is particularly strong. Like hyaline cartilage, there is collagen present, but here it is visible as distinct fibers. One major difference between this and hyaline cartilage is lack of a perichondrium.

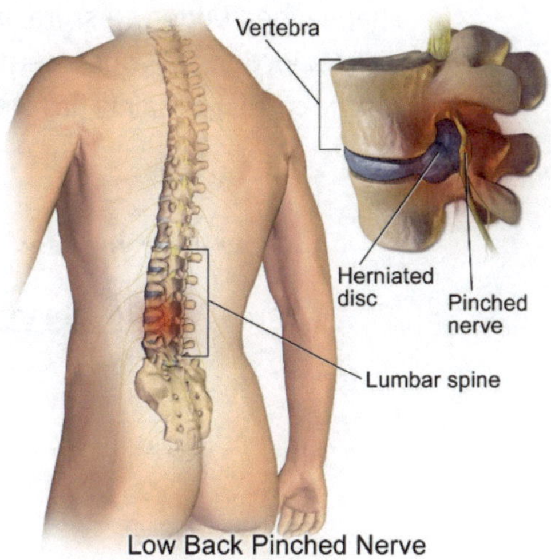

Low Back Pinched Nerve

The main cell type here is the chondrocyte which is surrounded by a **lacuna**. These chondrocytes are often bunched up in groups and form rows of cells (isogenous groups). You may also see fibroblasts that have spindle-shaped nuclei and are not surrounded by lacunae. They are also much rarer than chondrocytes, so you may have to look more closely to find them. Fibrocartilage is sometimes described as a mixture of dense regular connective tissue and cartilage, which is where it gets its name from.

**Draw what you see:**

<u>Make sure to label the following:</u> *chondrocytes, lacunae, fibroblasts (if present)*

Total Magnification of Drawing:



---

## Elastic Cartilage (16)

As the name suggests, elastic cartilage is composed of elastic fibers. This allows the tissue to change shape when under pressure, but return back to its normal shape when relaxed. A good example is the visible portions of your external ear (the auricle or pinna) and the tip of your nose. Like all cartilages, elastic cartilage mainly has chondrocytes encased in a lacuna. There is also a perichondrium as well.

**Draw what you see:**

Make sure to label the following: *perichondrium (dense IR), chondrocytes, lacunae*

Total Magnification of Drawing:



## Other Tissues:

This marks the end of connective tissue. So far we have covered all of connective tissue proper, and most of supportive connective tissue. Bones, blood, and lymph will be covered in their own chapters. Similarly, nervous and muscular tissues have their own sections.

## Review Questions:

1. Of the four major tissue types which two did we not study in this lab?
    1.
    2.
2. Fibers are a distinctive aspect of proper connective tissues. Fill out the table below to help you summarize these tissues.

| Tissue | Fibers Present | Describe the organization of the fibers |
|---|---|---|
| Dense Regular | | |
| Dense Irregular | | |
| Areolar | | |
| Adipose | | |
| Reticular | | |

3. Cartilage tends to be a confusing tissue for many students since they look similar at first glance. Fill out each blank in this table to help you compare the types of cartilage:

| Tissue | Major Cell Types Present | Locations in Body | Perichondrium Present? (yes/no) | Distinct, visible fibers (yes/no) |
|---|---|---|---|---|
| Hyaline | | | | |
| Fibrocartilage | | | | |
| Elastic | | | | |

4. What is particularly special about areolar connective tissue?

## Application Questions:

1. Identify each connective tissue below:

| | |
|---|---|
| 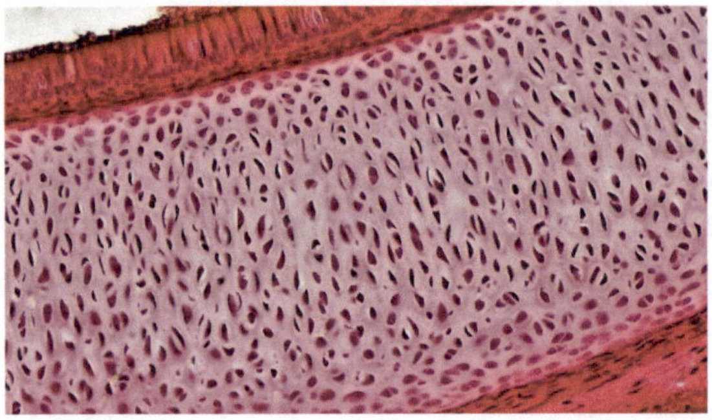 | 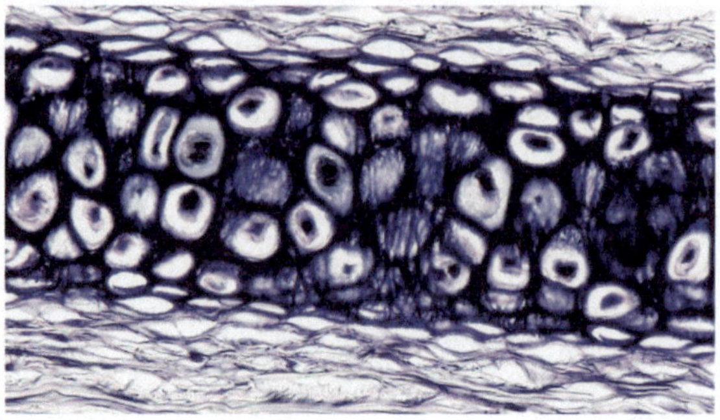 |
| | |
| 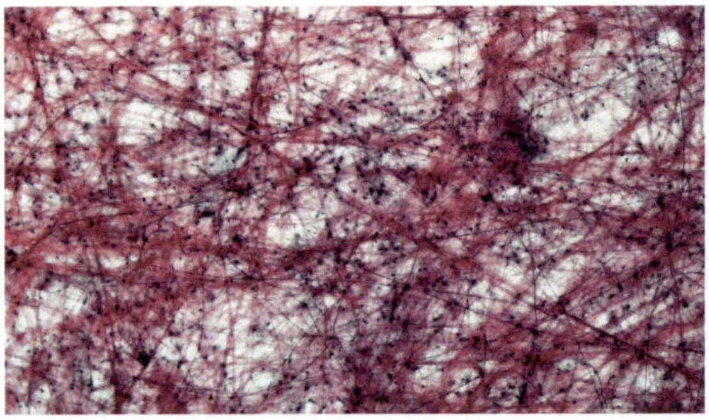 | 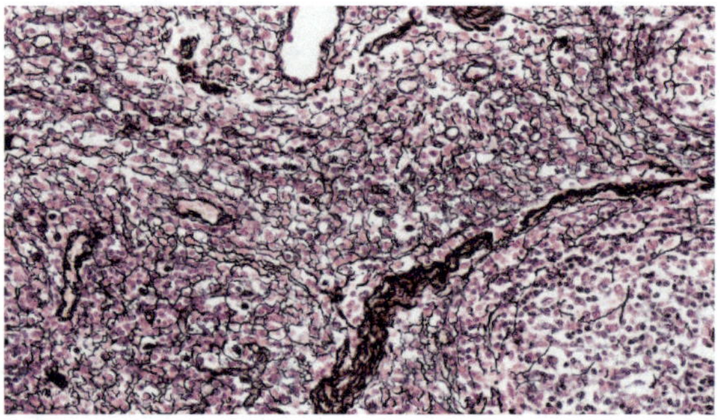 |
| | |

## Activity 2: Membranes and Glands

One major function of epithelial tissues is as **membrane** and **glands**. For the former, there are four types of membranes: cutaneous, synovial, mucous, and serous. Note that connective tissue comprises all of synovial membranes and part of cutaneous membranes. For the latter, epithelial glands have a shape, and a form of excretion (see below). Please note that merocrine excretion is also known as eccrine excretion. On the following pages you will review major concepts related to membranes and glands.

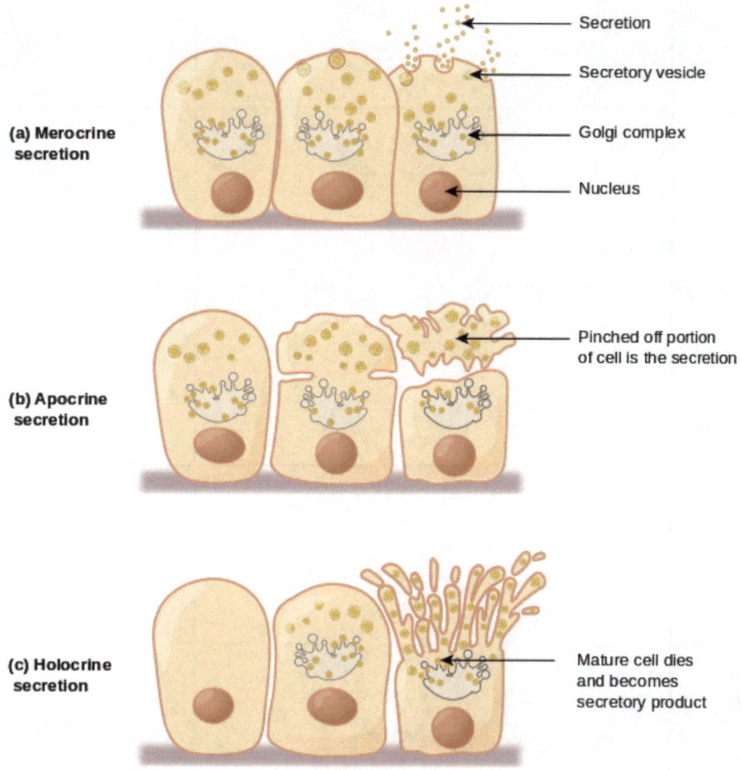

[6]

[6] "Types of Exocrine Excretion" by OpenStax is licensed under CC BY 4.0 International. Image cropped.

## Review Questions:

1. Make a quick drawing of all the various types of exocrine gland shapes and give an example/location. An example for simple tubular and simple coiled has been provided for you.

| Shape | Drawing | Example/Location |
|---|---|---|
| Simple Tubular |  | |
| Simple Coiled |  | |
| Simple Alveolar | | |
| Simple Branched Tubular | | |
| Simple Branched Alveolar | | |
| Compound Tubular | | |
| Compound Alveolar | | |
| Compound Tubuloalveolar | | |

2. List the 4 types of membranes and provide a brief description of their function.

## Application Questions:

1. Which surface (i.e., apical, basal, lateral, basolateral) is shed during apocrine excretion? How does this relate to its name?

2. Do holocrine cells survive after excretion? What major cellular process must occur to keep these kinds of glands running properly?

Intentionally Left Blank

4

# Lab 5

# The Integumentary System and Fascia

**Learning Objectives: Students will be able to...**

- Describe the overall organization of the integumentary system
- Label, describe, and identify important structures of hair
- Label, describe and identify important structures of the nail
- Describe the overall structure of an organism from skin down to bone

## Activity 1: Overview of the integumentary system

Skin is the largest organ in your body! Even though it is a large organ it is important to remember it is still composed of the 4 major tissue types previously discussed: epithelial, connective, nervous, and muscular. These 4 tissues working in concert can create enormously large organs such as skin. However, skin is but one of organ of the **integumentary system**. There are also **accessory structures** of the integumentary system to be aware of such as various glands, hair and nails.

Skin is organized into two major layers: the **epidermis and dermis**. Underlying the dermis is the **hypodermis**, also known as the **subcutaneous layer**. The hypodermis is composed of adipose tissue, but there is high content of intertwined collagen fibers. These fibers make up the **superficial fascia**. Due to high collagen content, the superficial fascia provides structural support for the body. Fascia is a hot field of study right now, so more information about its role(s) in the human body are still being researched.

Looking at the epidermis and dermis, we can split them into strata and layers, respectively. This would look like the following from superficial to deep:



1. Epidermis
    a. Stratum Corneum (oldest cells, thickest layer)
    b. Straum Lucidum
    c. Straum Granulosum
    d. Stratum Spinosum
    e. Stratum Basale (stem cells capable of generating new keratinocytes)
2. Dermis
    a. Papillary layer
    b. Reticular layer

The epidermis is 5 strata. However, all 5 of these strata combined make up stratified squamous epithelial tissue. In other words, we can ascribe names for the layers of cells found in stratified squamous tissue of the skin. The dermis is composed of connective tissue. Areolar for the papillary layer, and dense irregular for the reticular layer. Functionally, the epidermis waterproofs your skin. The major cell type found in the epidermis is the **keratinocyte**. These cells are packed with the protein keratin. Keratin is extremely hydrophobic, so it repels water. The connective tissue component of skin is much more complex, but one major cell type if fibroblast. It builds the extracellular matrix which contributes to the dermis's major function -giving skin its strength.

## Clinical Application: People of Color and Skin Disease

People with darker skin are much more likely to be diagnosed with late-stage skin cancer. The exact reason for this is unknown, but certainly the training that doctors receive focuses on white skin. Many skin diseases, including cancers, look very different between white and dark skin.

Medical professionals have been trying to address this disparity by producing textbooks aimed specifically at darker skin. This will ensure that the new generation of medical professionals will be equipped to deal with a diverse population of patients, and ensure the best health outcomes possible.

5

## Review Questions:

1. Imagine you are looking at the skin from the soles of your feet. What strata in the epidermis would you be looking at?

2. What strata is usualy the thickest

3. What is the pattern of movement you expect as epithelial cells of the straum basale mature?

4. Which strata from above is only found in thick skin?

5. What are the four major cells you find in the epidermis? What is the function of each?

6. What molecule provides waterproofing of skin?

7. What are the two layers of the dermis? Which is thickest? What type of tissue makes up each?

8. The hypodermis is primarily what type of tissue?

5

## Application Questions:

1. Provide the literal definition of the following words.

   a. Epidermis:________________________

   b. Hypodermis:________________________

   c. Arrector **pilli**:________________________

   d. Dermal papillae:________________________

   e. Sudoriferous (-ferous means bearing or having):________________________

5

## <u>Activity 2:</u> Integumentary system histology (17)

Draw a slide of hairy skin in detail below.

**Label the following:** epidermis, dermis, hair follicle, sebaceous gland, arrector pili muscle, and sweat gland.

Total Magnification:



## Application Questions:

1. Imagine this person had dandruff, what strata of skin would likely be enlarged and lack structural integrity?

## Activity 3: Human skin model

Locate the skin model in the classroom, and take it back to your desk for review. Try your best to identify each structure on your own.

In each box below, use a colored pencil to fill it in. On the next page, use the same colored pencil to color in each structure on the diagram. Closely compare the diagram and the model, note any difference you may see.

- Epidermis
- Dermis
- Hypodermis

} Label on the side of diagram, do not color in

- ☐ Meissner corpuscle
- ☐ Pacinian corpuscle
- ☐ Arrector pili muscle
- ☐ Free nerve endings
- ☐ Epidermal ridge
- ☐ Dermal papillae
- ☐ Hair root plexus
- ☐ Apocrine sweat gland (larger)
- ☐ Eccrine sweat gland (smaller)
- ☐ Sebaceous gland

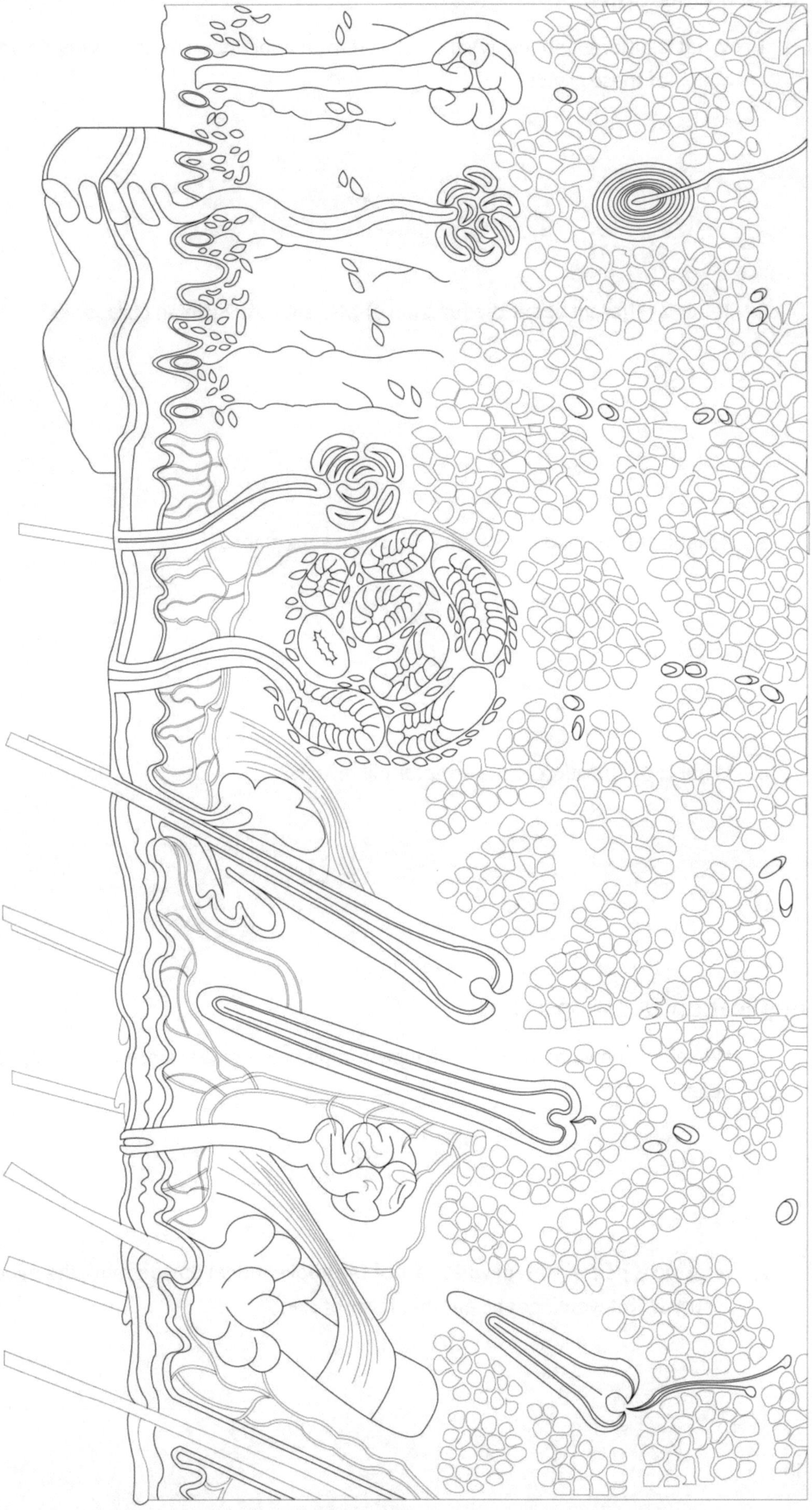

## Review Questions:

1. What is different about sweat produced from apocrine vs. eccrine sweat glands? What are the shape of these glands?

2. What is different about the function of Meissner vs. Pacinian corpuscles?

3. What is the method of excretion for sebaceous glands?

4. What is the function of the arrector pili muscle?



## Application Questions:

1. While not shown here, what cell type is implied to present since the hair on this model has color?

2. A tattoo is permanent. In what part of the above model do you think the needle penetrates? Justify why it would not be another layer of skin.

## Activity 4: Fingernail

Label each of the structures below:

- Nail body (used twice)
- Lateral nail edge
- Eponychium
- Proximal nail fold (used twice)
- Phalanx (bone)
- Hyponychium (used twice)
- Nail root
- Lunula (used twice)

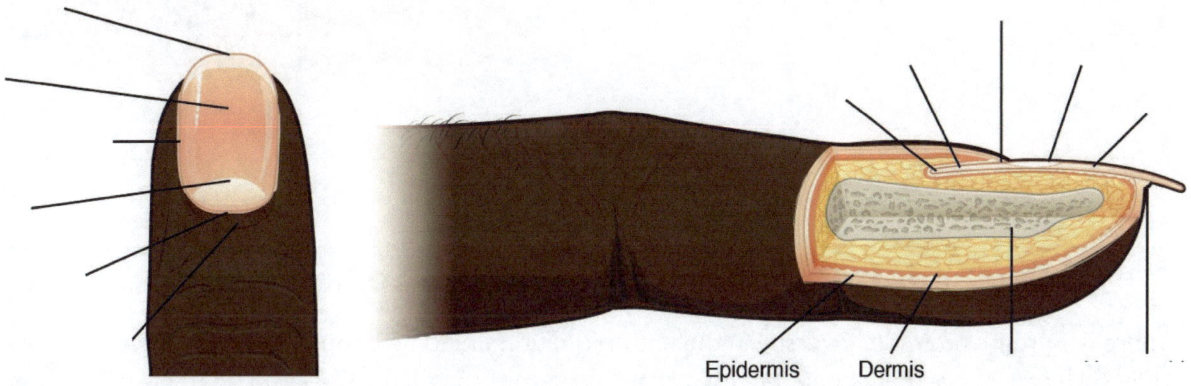


## Review Questions:

1. What protein is the major component of nails?

2. Why is the lunula white (or at least less red than the rest of the nail)?

---

[1] Access for free at https://openstax.org/books/anatomy-and-physiology/pages/1-introduction. Used by Permission. CC Attribution 4.0 International

## Activity 5: Hair

Label the following structures on the hair below:

- •Inner root sheath
- •Outer root sheath
- •Dermal papillae
- •Hair matrix
- •Sebaceous gland

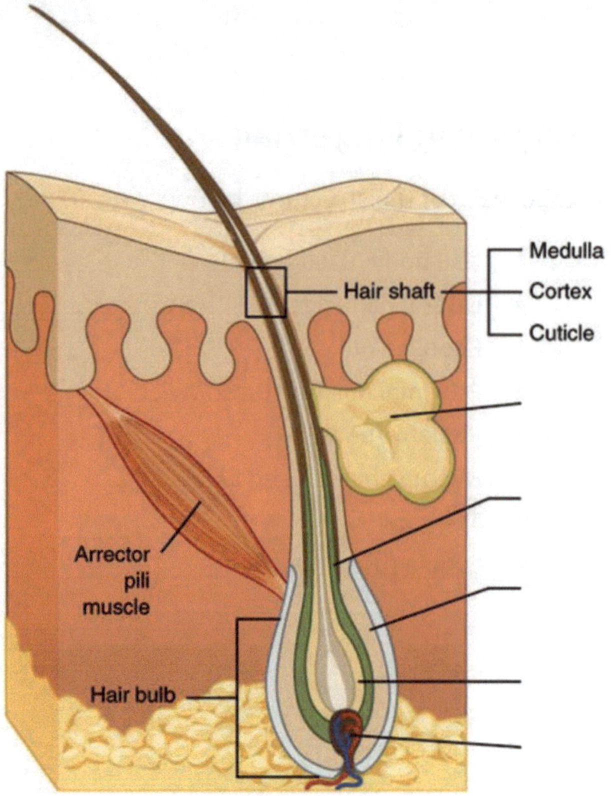


### Review Questions:

1. What is the function of dermal papillae?



### Every-body is Different: Curly Hair and Follicle Shape

There is a great amount of diversity in the nature of our hair. Some people have thick hair, some people things. Others have straight hair, but many have wavy and curly hair. The latter characteristic of hair is thanks to variabilities in the shape of the hair follicle. People with curly hair have a hooked-shaped hair follicle that almost looks like the letter 'C'. In cross-section, curly hair follicles have an oval shape (compared to straight hair which is circular). The cross-sectional shape along with the curved hair follicle causes the hair to grow in that pattern, creating the curls we see with our eyes.

[2] Access for free at https://openstax.org/books/anatomy-and-physiology/pages/1-introduction. Used by Permission. CC Attribution 4.0 International

## Application Questions:

1. Why would it be impossible for your hair to go white overnight?

## Activity 6: Chicken Wing Dissection

An organism is ultimately composed of multiple organ systems. In order to understand this concept, you will perform a dissection on a chicken wing. Just like the human upper limb, the chicken wing is organized in the same manner. In turn, you will once again use comparative anatomy to apply your knowledge to humans. This is a good point to mention the following statement:

**DISSECTION ITSELF IS NOT LEARNING! OBSERVATION AND UNDERSTANDING IS!**

This dissection involves dealing with relatively small structures that can easily be missed if you dissect too quickly. In turn, this dissection should be performed slowly and you will be given two chances to complete this dissection in your group. To start, grab the following materials:

- Scapel with blade
- Scissors
- Tweezers
- Gloves
- Dissecting tray with rubber mat
- Two chicken wings
- One 500ml beaker

The following will guide you along your dissection, **please follow the directions below carefully!**

1. In your group, choose one person who will not wear gloves or touch the chicken wing to answer questions in this packet and read steps out loud as you go along. **The chicken wing is raw meat, so it may have hazardous pathogens living on it!**
2. Set one wing aside for later, you will only work on one wing at a time.
3. Rotate the chicken wing until you fina a spot on the chicken wing that has bumps where feathers used to be ("goosebumps"). Depending on how the wing was butchered this could be anywhere, or maybe only in a small spot. **Stop here and answer the question below:**
4. Feathers and reptilian scales are thought to be homologous to hair in mammals (our common ancestor between reptiles/bird and mammals had something resembling these structures). With that in mind, what protein do you think makes up feathers?

| |
|---|
| |

5. Using tweezers (not fingers), grab a small piece of the skin and pull up with a significant amount of strength. Since the wing is light, someone will have to hold it with their fingers. As the skin is pulled up use a scalpel cut make a small incision **only** through the **skin and hypodermis. Make sure whoever is holding the chicken wing down has their fingers far out of the way of the blade!**
6. Slowly make your initial incision larger and larger while still pulling up on the skin with tweezers. As you do this you will notice the fascia of the chicken which will look like thin, white webbing. If

you see fat, you are cutting to superficially. If you are cutting into muscle, you are too deep. Stop here and answer the question below:

7. Of the 4 major tissue types, fascia is composed of which one? What fiber predominantly make up the webbing you are seeing?

8. Keep removing skin as in 6 until you have removed a large chunk and exposed the underlying muscle. Use this time to observe the fascia as you continue to pull and cut skin Stop and answer the following question before moving on.
9. Look at the underside of this chicken's skin, what specific tissue type is found on the underside? Is this the same or different from human integument?

10. Surrounding the muscle you will find a shiny tissue that consists of both the deep fascia and the epimysium (a protective sheet of dense irregular connective tissue that surrounds muscles). Cut through this tissue, the muscle will likely become dull and grainy after doing so.
11. Find a smaller muscle and attempt to trace it back to bone. Attached to the bone you will find a silvery piece of tissue called a tendon. Alternatively, pull on a small muscle to remove it from bone, the tendon will still be attached to muscle and will be a thin (~3mm) piece of tissue. Stop here and answer the following question.
12. What specific tissue type makes up a tendon?

13. Using scissors, start removing large pieces of muscle in order to expose the bone. As you do this, you may notice blood. Stop and answer the following question.
14. Arteries and veins are found in close proximity to muscles because they bring in what gas, and remove what gas? These gasses are related to what cellular process?

15. Once you are done removing all the muscles you should be left with bones, however they will not disarticulate (separate) without work. Holding them together are ligaments. Stop and answer the question below.
16. From what you have learned this far into this dissection, what do you think is the major difference between a tendon and a ligament? Considering their strength, what connective tissue do you think makes up a ligament? Some of these ligaments are made up of epithelial tissue as part of what type of membrane?

| |
|---|
| |

17. Using scissors cut ligaments to disarticulate the bones
18. Feel the spots that were articulating with other bones. Describe the texture of this tissue. What kind of cartilage do you think makes up these articular facets?

| |
|---|
| |

19. Take smallest long bone and place it a 500ml beaker. Fill the beaker with just enough white vingear to submerge the bone.
20. Place some saran wrap on top of the beaker and write a group name using a sharpie on the side of the beaker
21. Place the beaker on the tray
22. Repeat steps 1-18. Keep the same bone but this time wrap it with saran wrap and place it on top of your beaker. Do NOT submerge it in vinegar. Write your group name on the saran wrap so that it doesn't get misplaced.
23. Answer the questions in the section below before disposing of your chicken wing.



## Review Questions:

1. The bones that make up a chicken wing and the human upper limb. Below is chicken wing with labeled bones. Label the humerus, radius, and ulna on the human skeleton.

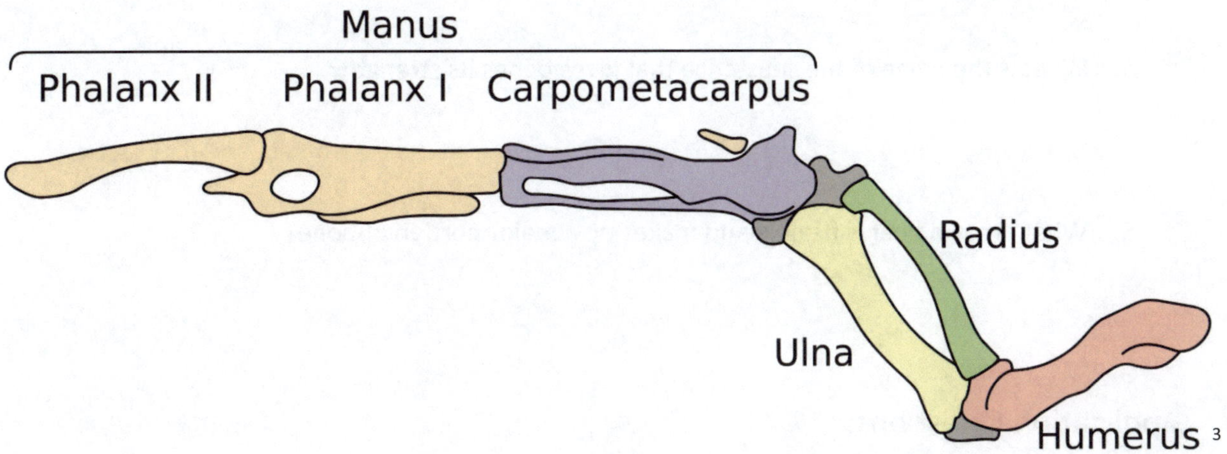


[3]

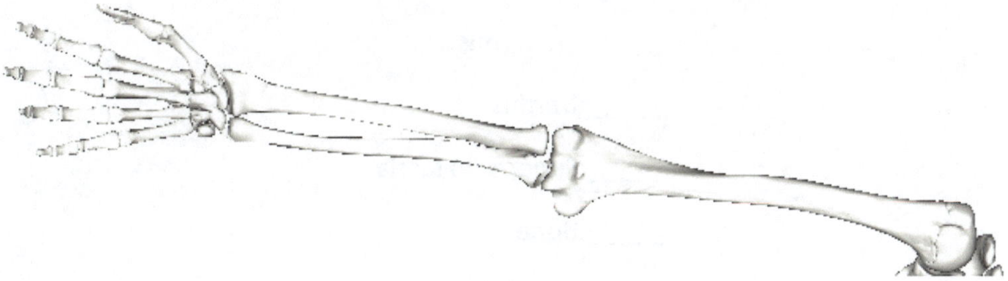

2. After labeling the above, what chicken bone did you place in the vinegar and saran wrap.

3. With the bones you have left, try to break one in half. Be careful of splinters, and keep the bone far from your face. Were you able to break the bone? What it easy or hard?

**For day 1 of this dissection skip the remaining review questions, and move on to the application questions section of this packet.**

[4] "Upper Limb" by BodyParts3D is licensed under CC BY-SA 2.1 Japan

4. What is the name of the substance that gives bones its strength?

5. What protein fiber is flexible and makes up a major portion of bone?

## Application Questions:

1. Place the following layers in order from most superficial to deep.

______:Deep Fascia

______:Epidermis

______:Dermis

______:Superficial fascia

______:Bone

______:Muscle

The remaining application questions are to be answered the following class period. For now, do the following to clean up:

1. **Discard of any solid tissue in the appropriate biohazard bin**
2. **Rinse all tools and trays**
3. **Dry the tools and trays with paper towels and place them back in the correct area**
4. **Spray your bench with disinfectant and wipe it down with a paper towel**

2. Of the two substances you identified in review questions 2 and 3, which do you think dissolved in the vinegar? What substance is left over then?

3. What does the above tell you about how the bone needs to balance and role of these two substances?

4. Teeth are also composed of the substance you identified in review question 2. What do acidic substances such as sodas do to your teeth?

5

Intentionally Left Blank

5

# Lab 6

# Osseous Tissue and the Skeletal System

## Learning Objectives: Students will be able to…

- label and list parts of bone both at the gross and microscopic level
- Compare and contrast the axial vs. appendicular skeleton
- Identify bones and bony landmarks
- Describe which bones articulate with which, and the number of each bone present
- Describe the layout of the vertebral column (and the number of each vertebrae)
- Describe the layout of the ribcage

## Exercise 1: Bone histology

Bone is the foundation by which your muscles connect to your body and allow for movement. Of course, the skeleton alone provides enormous structural support. The skeleton is composed of connective tissue. Specifically, there are two types of bone (osseous) tissue: **compact** and **spongy**. The latter is also sometimes known as cancellous bone. For compact bone, there is dense packing of the extracellular matrix which gives this type of bone tissue enormous strength. Spongy bone, on the other hand, is not as strong, but has greater flexibility and space for other substances (e.g., bone marrow).

The major structural molecule of bone is **hydroxyapatite**. The chemical formula of which is $Ca_{10}(PO_4)_6(OH)_2$. The calcium ('Ca') is hydroxyapatite is why so many commercial and food products push the benefits of calcium for strong bones. However, be aware that your bones are not "made of calcium". They are made of hydroxyapatite, a complex crystalline molecule that is much more complex than an atom of calcium. Your bones do not start off with hydroxyapatite though. Through a process called **ossification**, they develop the hydroxyapatite that gives them strength.

### Clinical Application: Vitamin D Deficiency

Parlaying off of the previous section on the integumentary system, The UV rays from sun aid in the chemical reactions necessary for our body to synthesize Vitamin D. This is one reason, among many, why going outside can be a very healthy thing to do. Vitamin D plays a critical role in our body's ability to absorb and utilize calcium. Severe Vitamin D deficiency can cause bones to become brittle.

One major issue is that people of color have skin that does not absorb Vitamin D well. Additionally, those whose cultural or religious beliefs require being fully-enshrouded while outside may have trouble producing enough Vitamin D. Luckily, Vitamin D can easily be supplemented via pills or diet.

View the following slides on the microscope

## Compact bone (18)

The functional unit of compact bone is the **osteon**, a circular structure consisting of smaller **concentric lamellae**. At the center of the osteon you have the **central canal**, a hollow structure containing blood vessels that bring nutrients to the ossified bone. To aid in this, small **canaliculi** ("tiny canals") penetrate outwards from the central canal and form pores that allow for the embedded **osteocytes** (embedded in lacunae) to be bathed in nutrients.

In between osteons you also have some structures such as **interstitial lamellae** and surrounding the bone as a whole you have the **circumferential lamellae**. There are also structures that you are unlikely to see such as **perforating canals** which link the central canal veins and arteries of adjacent osteons together. Similarly, there are also likely **osteoblasts** and **osteoclasts** somewhere in the bone this tissue was extracted from, but you won't see them on this slide.

**Draw what you see**

**Label the following:** Osteon, central canal, concentric lamellae, osteocytes inside of lacunae, canaliculi, interstitial lamellae

Total Magnification:

## Spongy (cancellous) bone (19)

Slightly different from compact bone, spongy bone does not have osteons or a central canal. Its thinner structure means that no central canal is necessary. Despite that, there are still canaliculi present to allow deeper layers of lamellae to have blood brought to their osteocytes. Another feature of spongy bone that is different from compact bone is an abundance of osteoblasts. These cells which are not embedded in matrix are actively remodeling bone.

As you may be aware, spongy bone contains bone marrow. For spongy bone in epiphyses you will likely see red bone marrow (where B-cells and red blood cells mature). For spongy bone in the medullary cavity, you will see yellow bone marrow

**Draw what you see**

**Label the following:** Osteocytes, osteoblasts, bone marrow, trabeculae

Total Magnification:

## Review Questions:

1. What is the purpose of canaliculi?

2. Fill out this table comparing compact and spongy bone:

| Tissue | Osteons? | Lamellae? | Trabeculae? | Cell Types Present... |
|---|---|---|---|---|
| Compact Bone | | | | |
| Spongy Bone | | | | |

6

# Exercise 2: Major bones of the human body

The human skeleton is composed of 2 major divisions: **the axial skeleton**, and **the appendicular skeleton**. The former consists of bones on the midline of the body including the vertebral column, ribcage, and cranium. The latter is then every bone hanging off the midline including those that make up your upper limb, lower limb, and pelvic girdle.

In total, the human body has 206 bones, but to make matters more complex each bone has bony landmarks, areas of the bone with certain names. Often, these bony landmarks are important sites for muscle attachment. In turn, it will be important to master them before we move on the muscular system.

Before that, you must learn the names of the major human bones. The image to the right serves a starting point, and covers some of the most important bones.

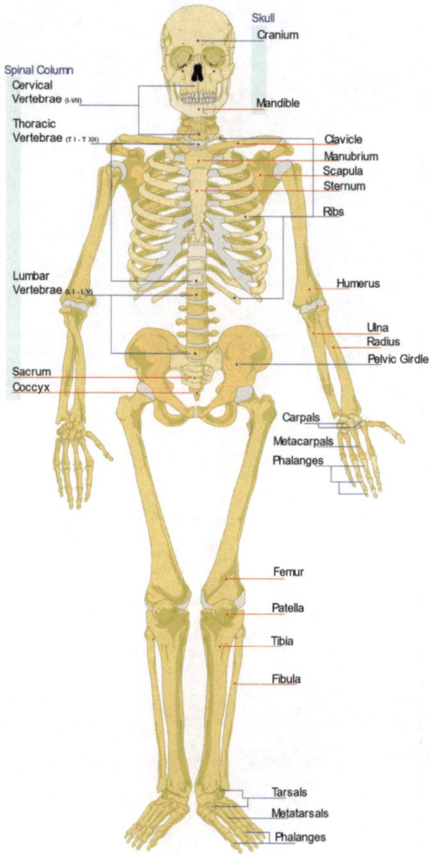




[1] Courtesy of pixabay.

## Review Questions:

1. From the images below, identify each disarticulated bone:

| | | |
|---|---|---|
|  |  |  |
| | | |
| 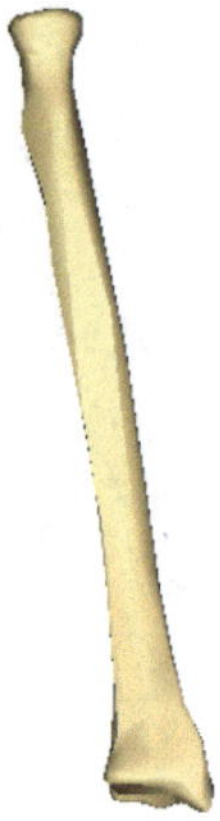 |  | 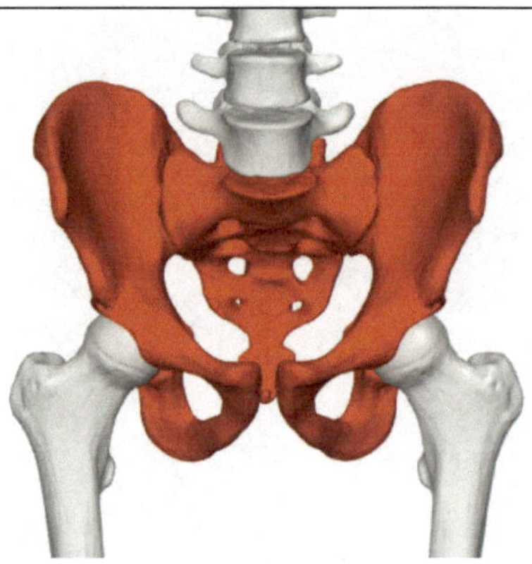 (all bones together) |
| | | |

2. For each question below, answer what bone articulates with the one listed based on the direction terms.

   a. The humerus articulates...

      i. Proximally with the ________________

      ii. Distally with the ________________ and ________________

   b. The femur articulates...

      i. Proximally with the ________________

      ii. Distally with the ________________

   c. The *os coxae* ("hip bone") articulates...

      i. Anteriorly with the other ________________

      ii. Posteriorally with the ________________

      iii. Laterally with the ________________

   d. The tibia articulates...

      i. Laterally with the________________

      ii. Distally with the ________________

3. Which bones from the ones you assembled belong to the axial skeleton?

6

4. The "pelvic girdle" contains what bones?

## Application questions:

1. In class we learned about 7 types of bones, give an example of the ones listed below. **Try this without looking up the answers in your book.**

| Bone Type | Example |
|---|---|
| Long Bone | |
| Short Bone | |
| Sesamoid Bone | |
| Irregular bone | |
| Flat bone | |

2. The sternum and rib cage are incredible meshwork of bones. You have almost as many bones here as you do in your entire vertebral column! Why do you think humans have evolved to have so many bones in this general location?

3. The fibula is a non-weight bearing bone. Explain why this is true based on the articulations you observe.

## Exercise 3: The Vertebral column

In the last exercise, you assembled most bones of the human body besides for the vertebral column. Now you will have an opportunity to do this. The vertebral column is composed of about 32 bones total (24 presacral, 5 fused sacral, and 3-5 fused coccygeal). Each vertebra has its own profile which makes most easy to identify. Use the below table to help with the exercise on the next page.

| | | |
|---|---|---|
| Cervical Vertebrae<br>(7 total) | $C_1$ (Atlas) | 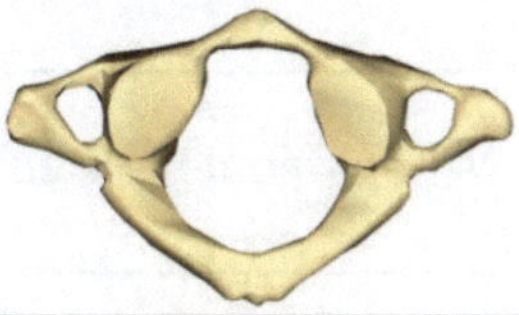 |
| | $C_2$ (Axis) | 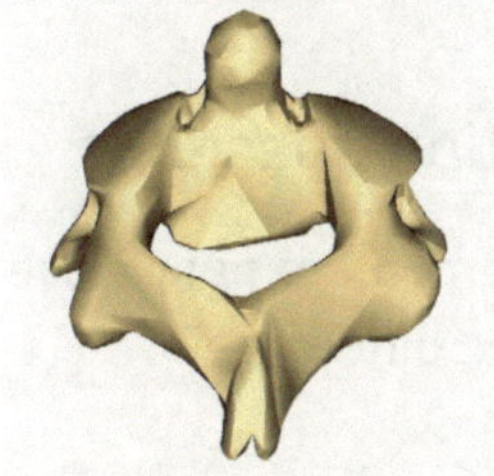 |
| | $C_3$-$C_7$ | 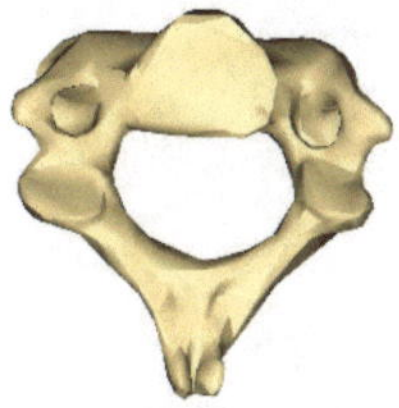 |
| Thoracic Vertebrae<br>(12 total) | $T_1$-$T_{12}$ | 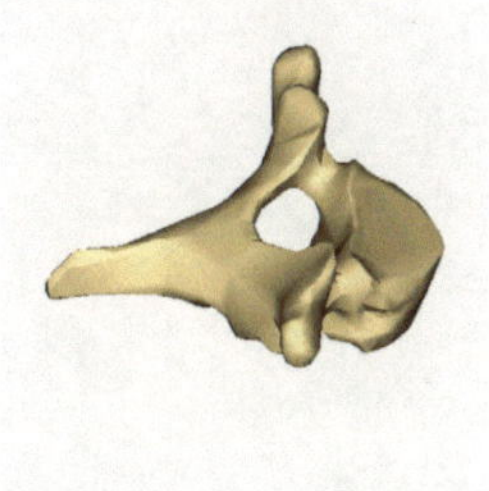 |
| Lumbar Vertebrae<br>(5 total) | $L_1$-$L_5$ | 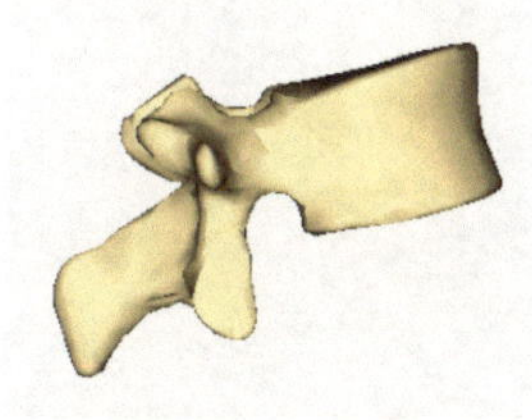 |





6

## Review questions:

1. When people talk about "neck pain" they are talking about pain near their _________________ vertebrae.
2. When people talk about "lower back pain" they are talking about pain near their _________________ vertebrae.
3. When people talk about "upper back pain" they are talking about pain near their _________________ vertebrae.
4. When people talk about "tail bone pain" they are talking about pain in their _________________ bone.

## Application questions:

1. Sometimes people use the words "spinal cord" and "vertebral column" interchangeably. What is the difference?

## Exercise 4: Bony landmarks of the axial skeleton

As described above, bony landmarks (AKA bony regions or features) are usually the site of muscle attachment or have some other unique property (e.g., a nerve or vein passes through them). It will be important to learn these as we go forward since these structures will be referenced again-and-again.

As you work through these next pages you will be coloring (and optionally numbering) each bony landmark. Color each box to the left of the bone or landmark and use the same color to color in the drawing. Add numbers if you wish. Do not label the complete name directly on the diagram, so that you can study from these pages later without seeing the answers. Using your bone box here will also be helpful for studying since you can manipulate the bones.

*Cranial Bones (external view):*

- ☐ Frontal bone
- ☐ Parietal bone (2x)
- ☐ Occipital bone
- ☐ Temporal bone (2x)
- ☐ Sphenoid (eye socket too!)
- ☐ Ethmoid (barely visible

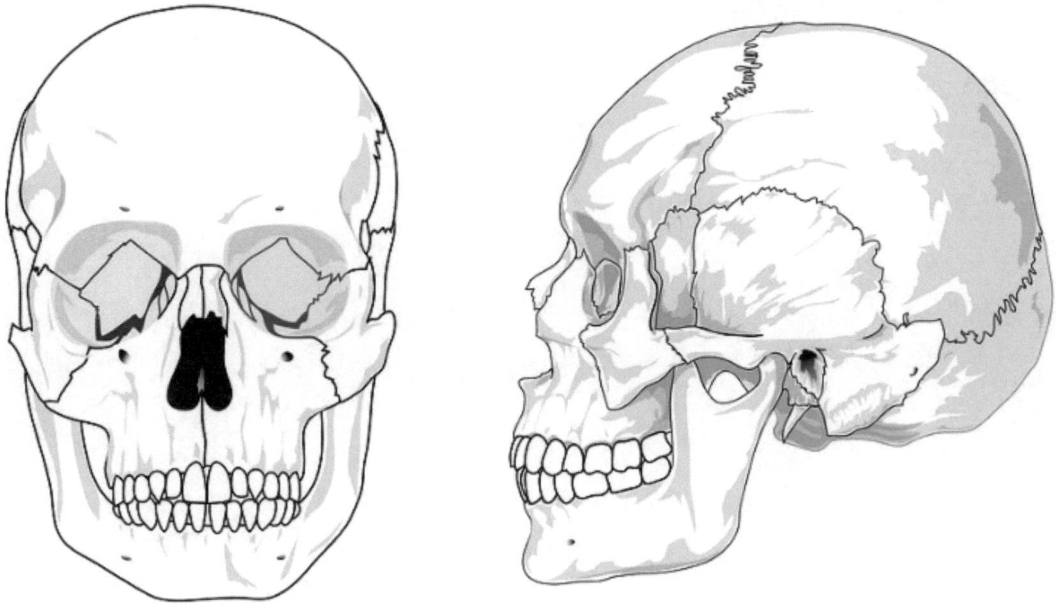

*Cranial Bones (sutures):*

- ☐ Coronal suture
- ☐ Saggital suture
- ☐ Lambdoid suture
- ☐ Squamous suture

*Temporal bone (landmarks):*

- ☐ Zygomatic process
- ☐ Styloid process
- ☐ Mastoid process
- ☐ External acoustic meatus
- ☐ Squamous portion
- ☐ Petromastoid portion

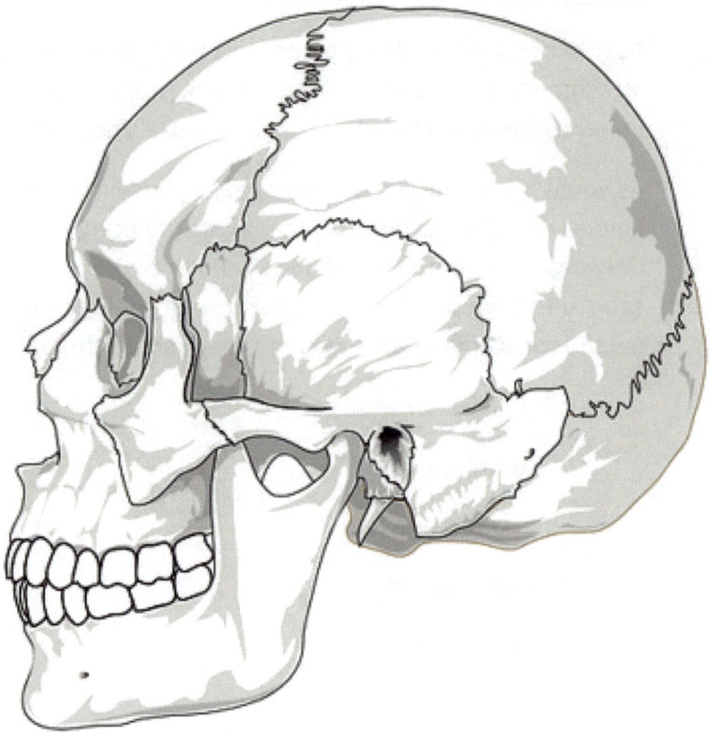

6

*Sphenoid bone*

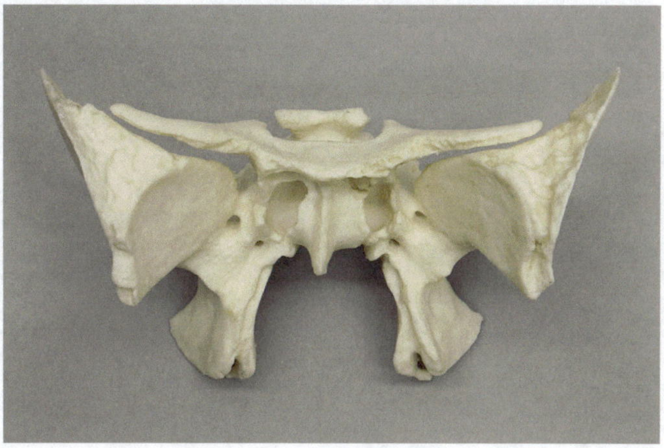

- [ ] Optic canal
- [ ] Greater wing
- [ ] Lesser wing
- [ ] Foramen ovale
- [ ] Foramen spinosum
- [ ] Foramen rotundum (hard to see)
- [ ] Sella turcica (incl. hypophyseal fossa)

*Sphenoid bone (articulated, anterior view):*

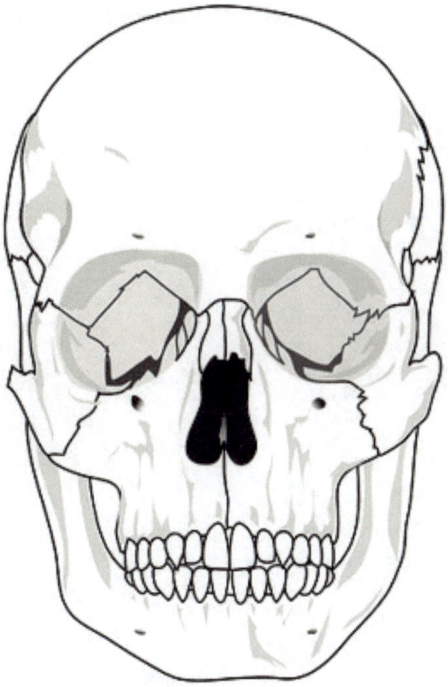

- [ ] Superior orbital fissure (not drawn well here)
- [ ] Sphenoid bone

6

[2] "Bone Clones cast of a human sphenoid bone" by Bone Clones is licensed under CC BY-SA 3.0 Unported

*Sphenoid bone (articulated, inferior view of skull):*

- ☐ Greater wing
- ☐ Foramen ovale

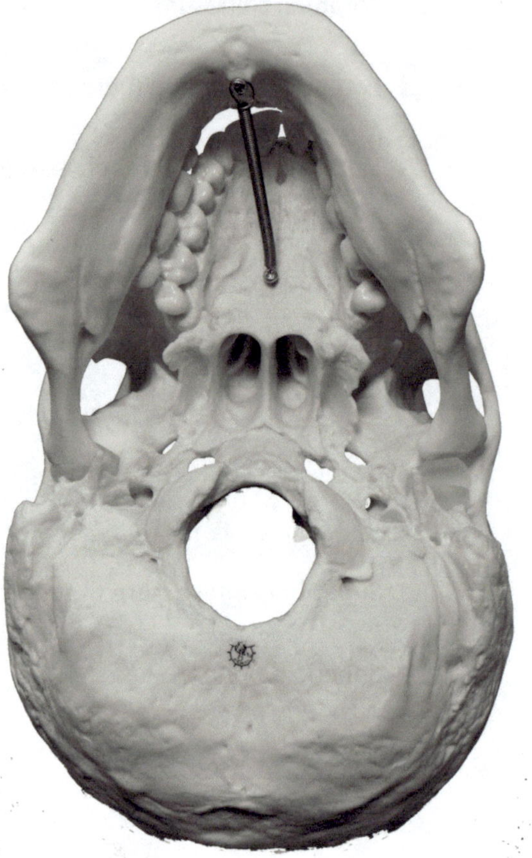

6

[3] "Skull--Inferior projection." by Bruce Forciea is licensed under CC BY-SA 4.0 International

*Ethmoid bone (disarticulated, front view):*

- ☐ Crista galli
- ☐ Perpendicular plate
- ☐ Ethmoidal air cells
- ☐ Middle nasal concha
- ☐ Cribiform plate (slight visible)

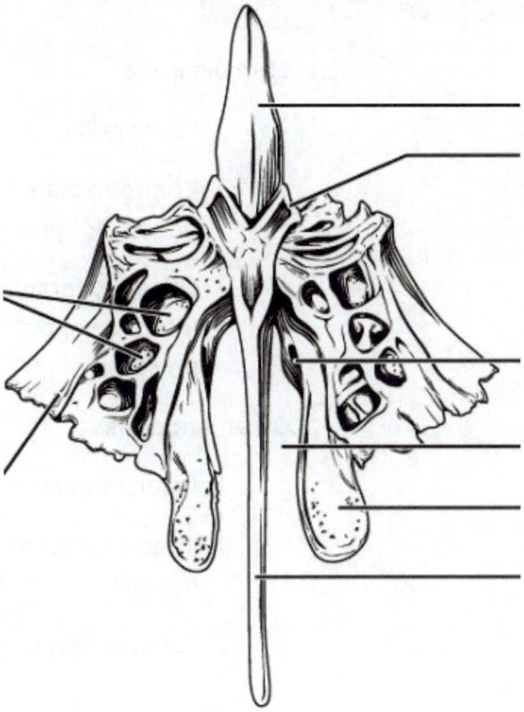

*Ethmoid bone (superior view, disarticulated)*

- ☐ Cribiform plate
- ☐ Olfactory foramina

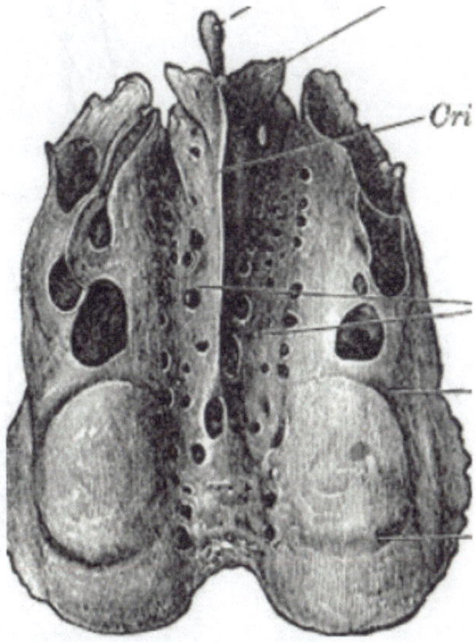



[4] "Ethmoid Bone" by OpenStax is licensed under CC BY 3.0. Image cropped.

*Skull (interior, superior view):*

- ☐ **Ethmoid bone**
  - ☐ Crista galli
  - ☐ Cribiform plate (not drawn, but present)
  - ☐ Olfactory foramina (not drawn, but present)
- ☐ **Other landmarks**
  - ☐ Jugular foramen
  - ☐ Internal acoustic meatus
  - ☐ Foramen lacerum
  - ☐ Hypoglossal canal
- ☐ **Sphenoid bone**
  - ☐ Greater wing
  - ☐ Lesser wing
  - ☐ Foramen rotundum
  - ☐ Foramen ovale
  - ☐ Foramen spinosum
  - ☐ Optic canal/foramen
  - ☐ Sella turcica

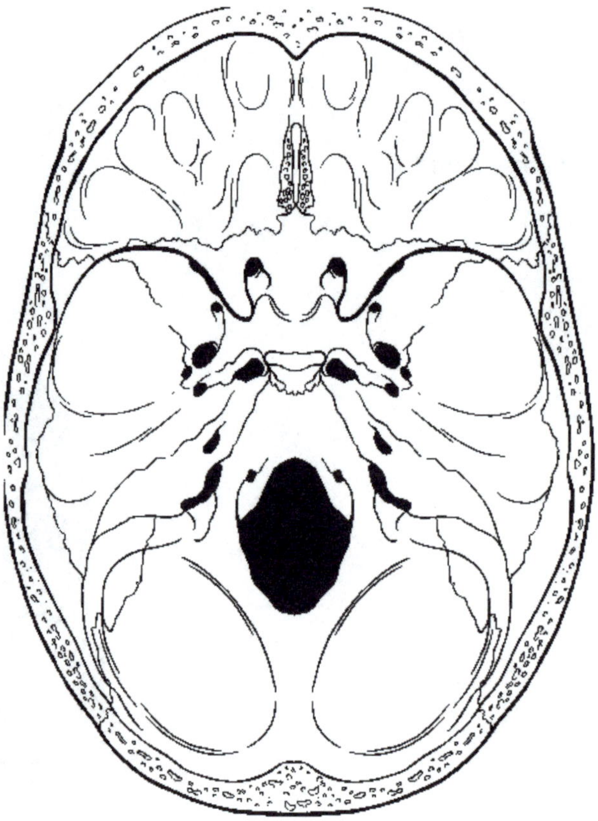

*Sinuses of the skull (coronal and sagittal, MRI, inverted colors):*

- ☐ Frontal sinus
- ☐ Ethmoid air cells/sinus
- ☐ Maxillary sinus (coronal only)
- ☐ Nasal cavity (not a sinus!)
- ☐ Sphenoidal sinus (sagittal only)

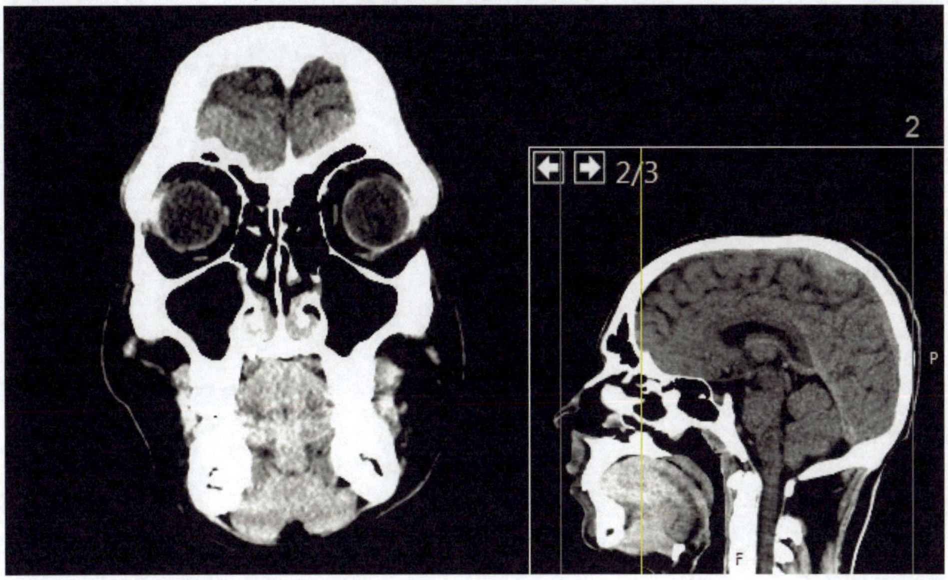


*Sinuses of the skull (sagittal):*

- ☐ Frontal sinus
- ☐ Sphenoidal sinus
- ☐ Sella turcica (not a sinus)

(other sinuses not shown)

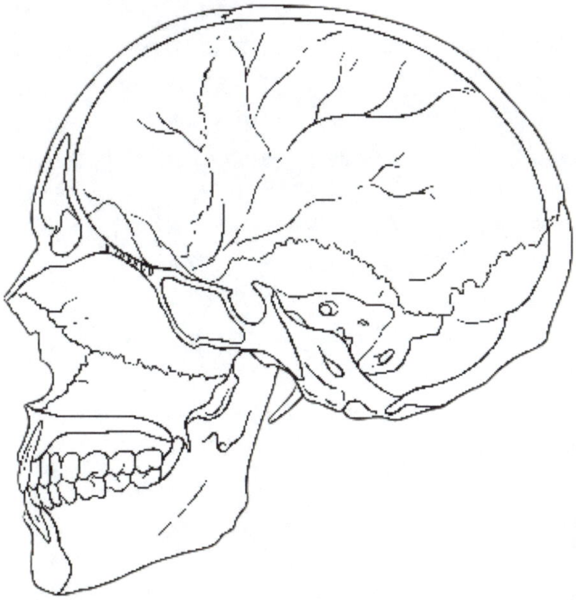



[6] "Skull human sagittal section" by Patrick J. Lynch is licensed under CC BY 2.5 Generic

*Facial bones:*

- ☐ Zygomatic (2x)
- ☐ Maxillae (2x)
- ☐ Mandible
- ☐ Vomer (barely visible)
- ☐ Lacrimal (barely visible, 2 present)
- ☐ Nasal (2x)

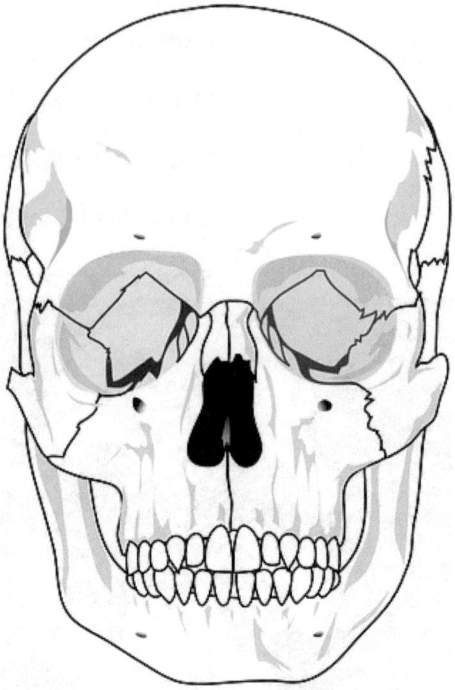

(Palatine and inferior nasal concha bones not shown)

Major foramen of frontal view:

- ☐ Supraorbital foramen (or notch)
- ☐ Infraorbital foramen
- ☐ Mental foramen

6

*Inferior view bones and landmarks:*

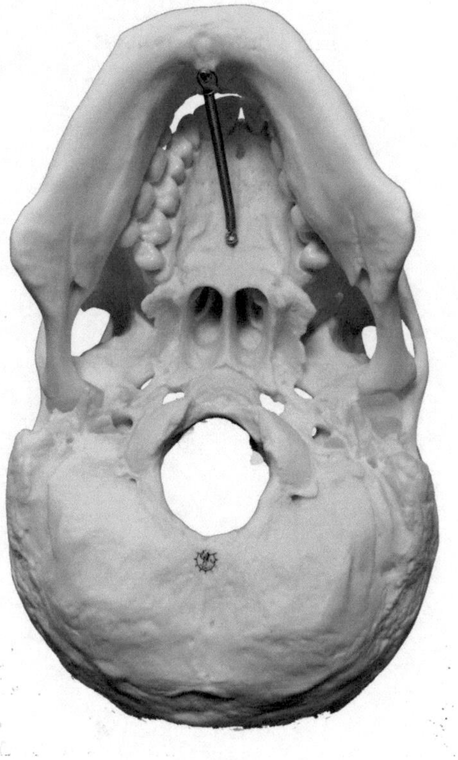

- ☐ Palatine bone
- ☐ Foramen magnum
- ☐ Occipital condyle
- ☐ External occipital protuberance
- ☐ Carotid canal
- ☐ Jugular fossa
- ☐ Hypoglossal canal (not shown)

*Bony landmarks of mandible:*

- ☐ Ramus (of mandible)
- ☐ Coronoid process (of mandible)
- ☐ Condylar process/Mandibular condyle
- ☐ Genu

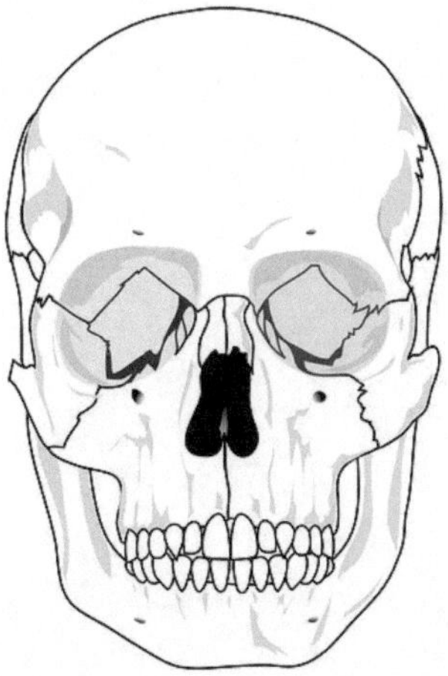

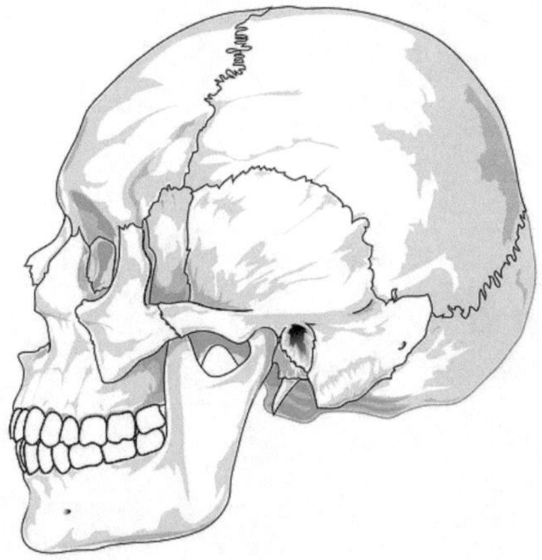

6

*Associated bones:*

- ☐ Auditory ossicles
  - ☐ Malleus
  - ☐ Incus
  - ☐ Stapes

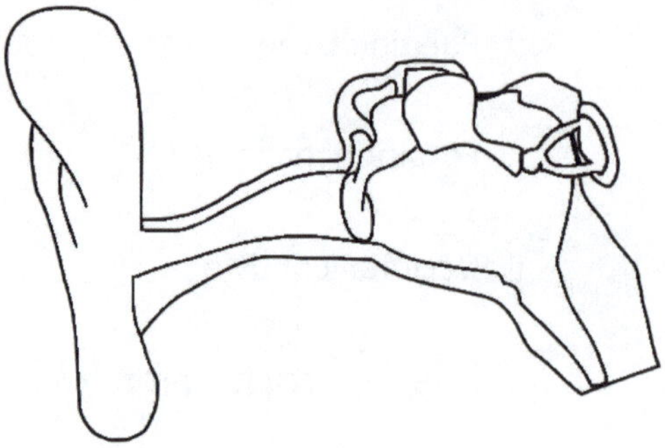

- ☐ Hyoid bone

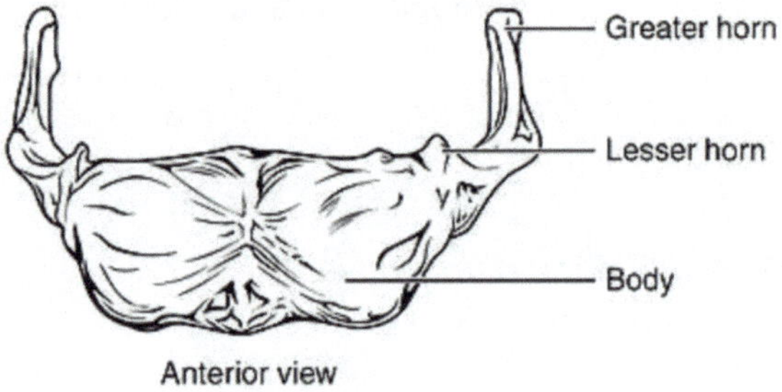


6

[8] "Hyoid Bone" by OpenStax is licensed under CC BY 3.0. Image cropped.

*Spinal cord curvatures:*

- ☐ Cervical (2°)
- ☐ Thoracic (1°)
- ☐ Lumbar (2°)
- ☐ Sacral (1°)

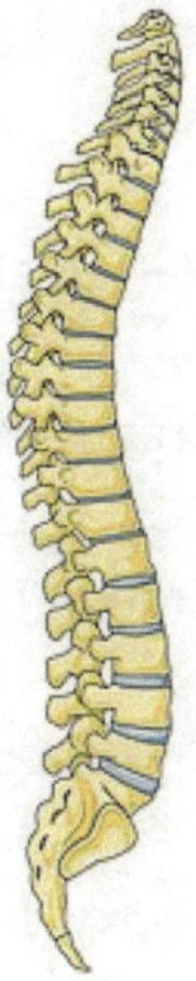

*Vertebrae:*

- ☐ Body
- ☐ Pedicle
- ☐ Lamina
- ☐ Superior and Inferior Articular Process
- ☐ Transverse foramen (cervical only)
- ☐ Spinous process
- ☐ Vertebral foramen
- ☐ Articular facet for rib (thoracic only)
- ☐ Transverse process

| Cervical (C3-C7) | Thoracic | Lumbar |
|---|---|---|
|  | 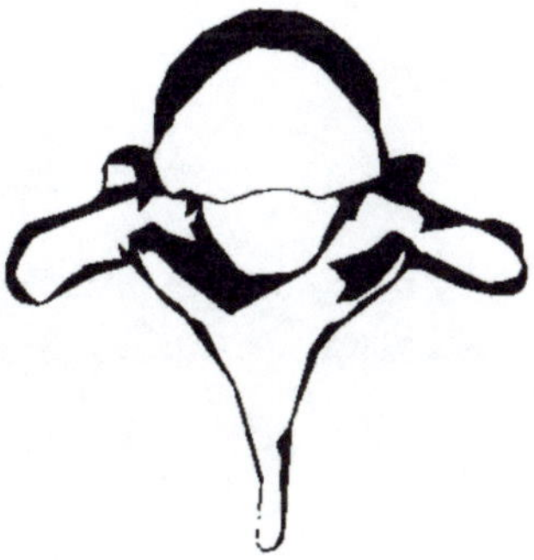 | 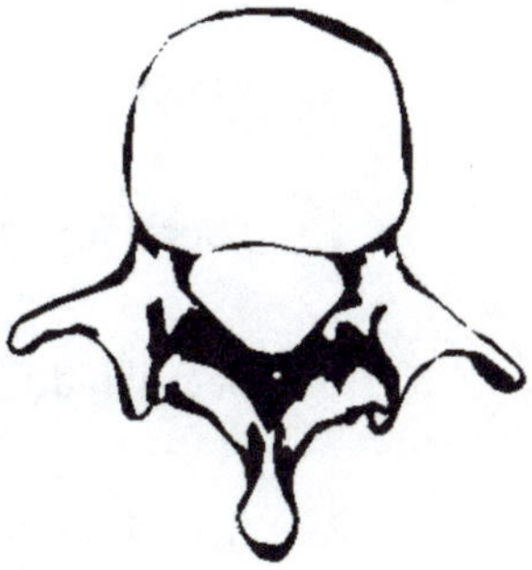 |
| 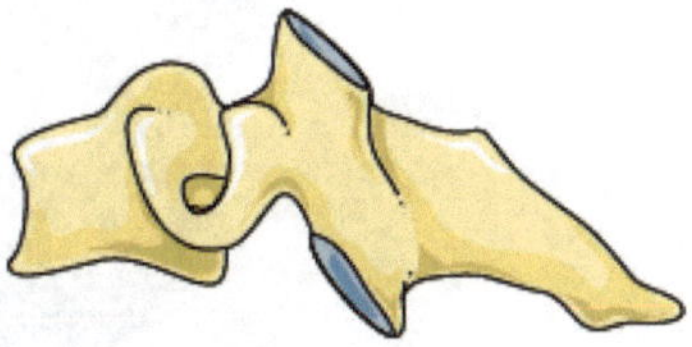 | 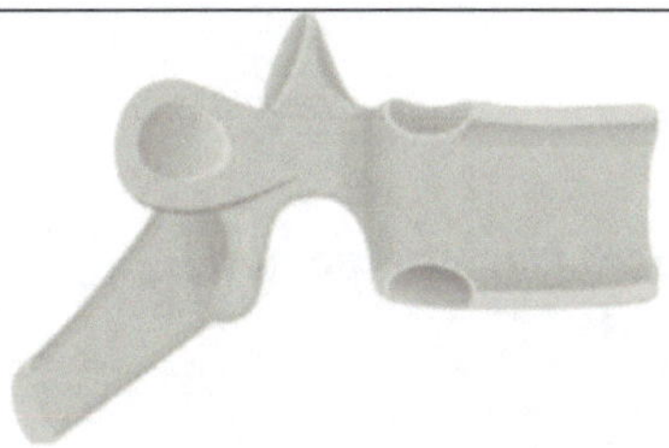 |  |

*Atlas and Axis:*

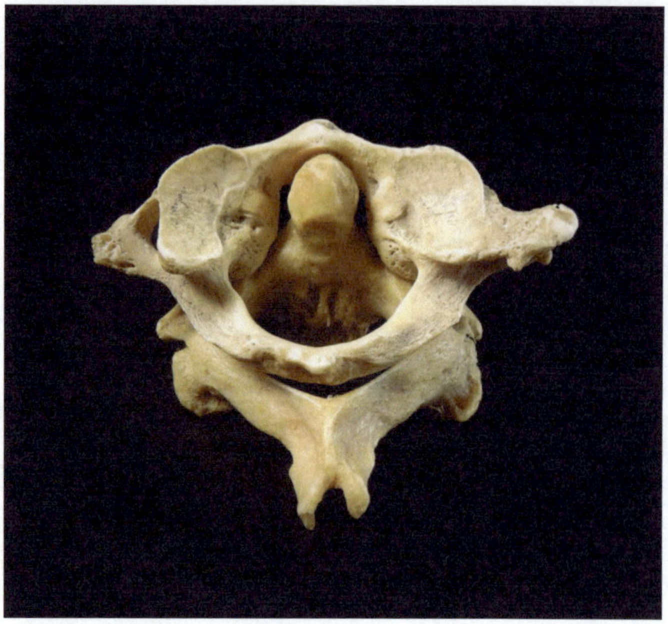

- ☐ Vertebral foramen
- ☐ Transverse foramen
- ☐ Transverse process
- ☐ Spinous process
- ☐ Dens
- ☐ Superior articular facets

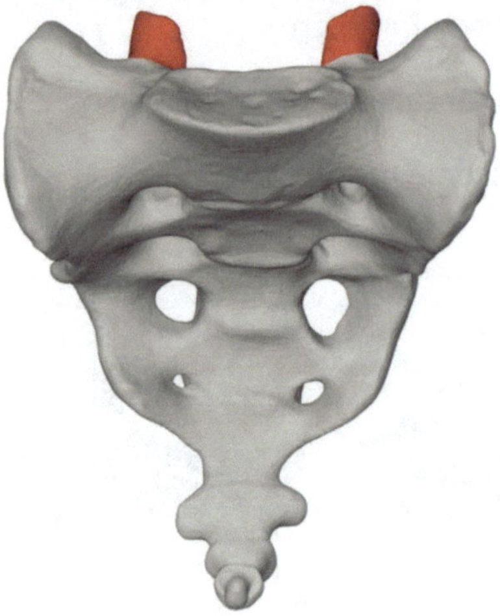

*Sacrum and coccyx:*

- ☐ Sacrum (outline with color)
- ☐ Coccyx
- ☐ Sacral foramina
- ☐ Sacral hiatus
- ☐ Sacral canal
- ☐ Ala
- ☐ Articular surface
- ☐ Sacral foramina (anterior and posterior)

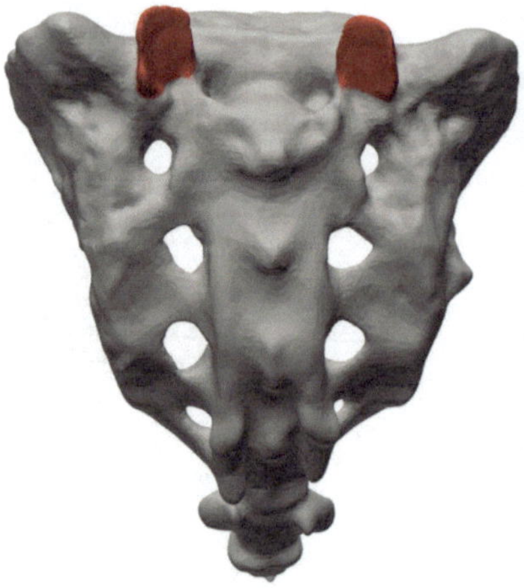

6

*Sternum:*

- ☐ Body
- ☐ Xiphoid process
- ☐ Manubrium
- ☐ Jugular notch
- ☐ Sternal angle

*Rib Cage:*

- ☐ True ribs
- ☐ False ribs
- ☐ Floating ribs
- ☐ Angle (choose 1 rib)
- ☐ Tubercle (not shown)
- ☐ Head and neck (not shown)
- ☐ Costal cartilage (not bone)
- ☐ Sternum (separate bone)

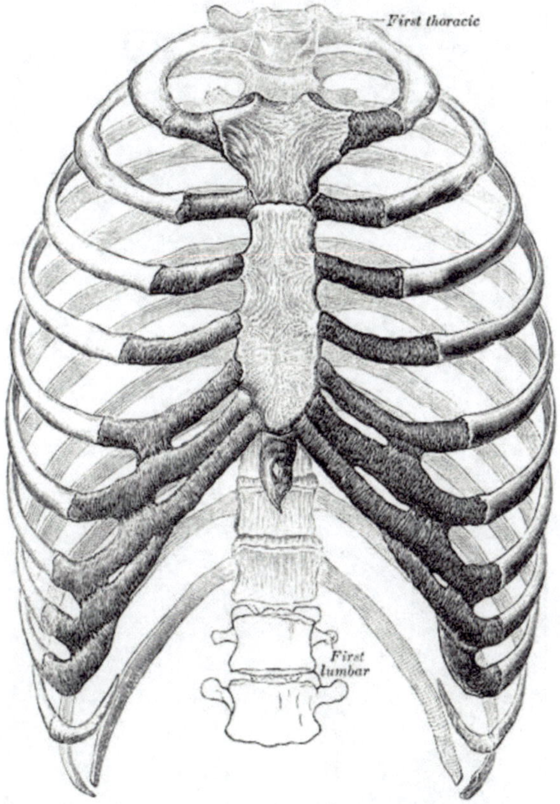


6

---

## Review questions:

Identify each of the bony landmarks below.

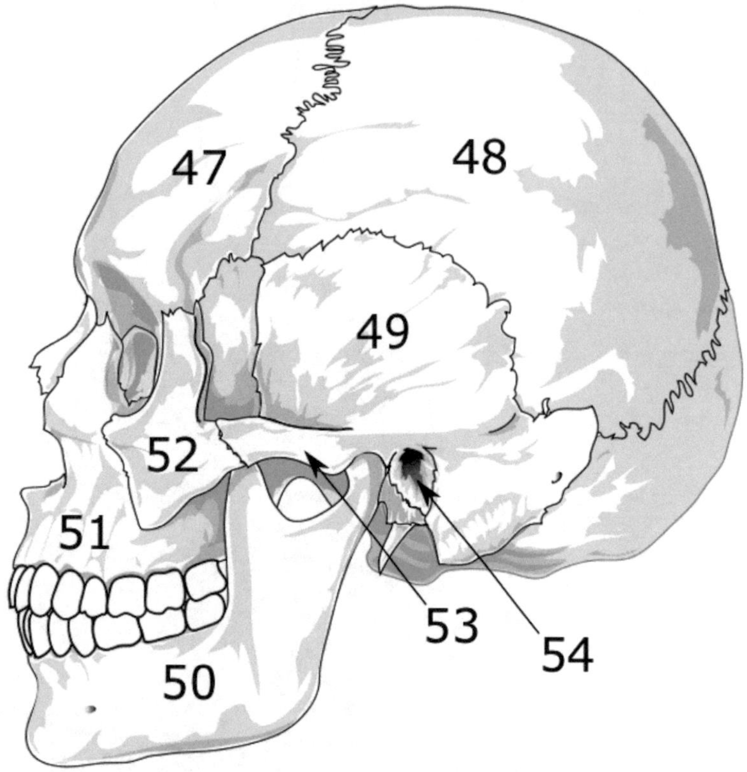


47.

48.

49.

50.

51.

52.

53.

54.

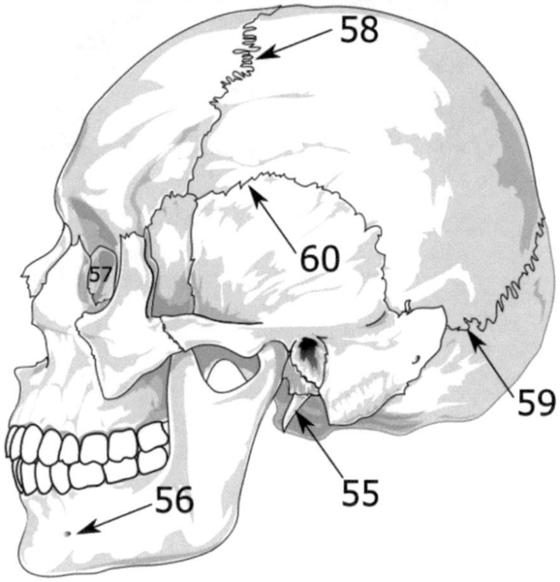


55.

56.

57.

58.

59.

60.

| | |
|---|---|
| 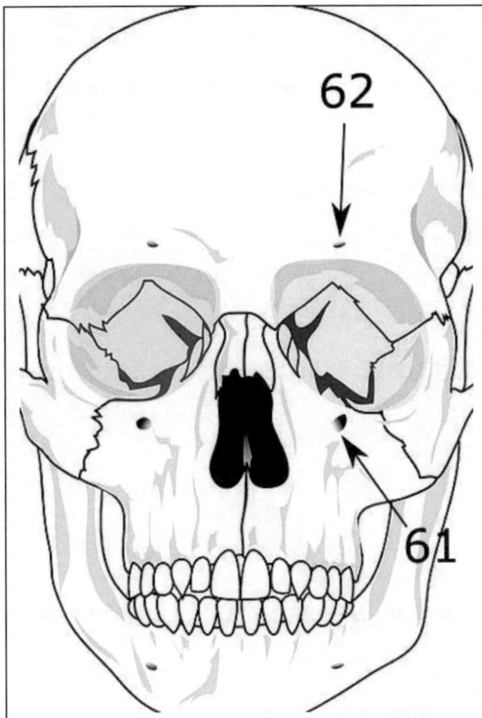 | 61<br>62 |
| 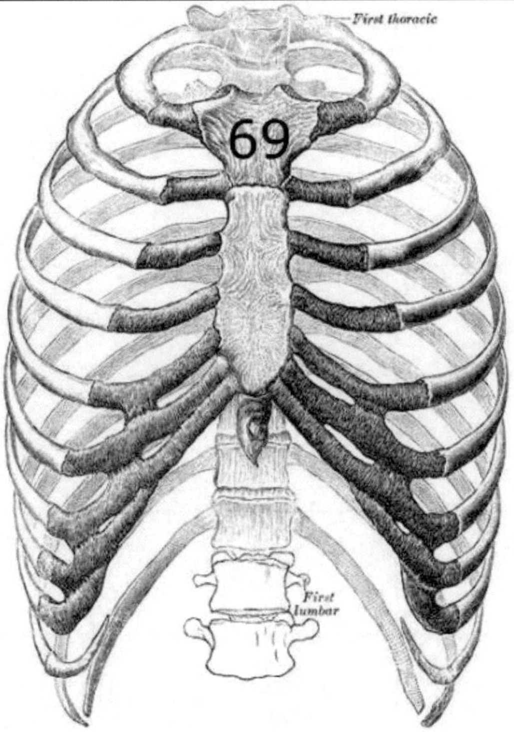 | 69. |

6

## Application questions:

1. The pre-sacral vertebral column has some landmarks found in all three vertebrae types, what are they?

2. The eye orbit ("socket") itself is not a bone, it is actually composed of seven bones. List them.

3. What passes through each of the following foramen? You will need your book, the internet, or your own knowledge to answer this question.

| Foramen | What passes through? |
|---|---|
| Magnum | |
| Supraorbital | |
| Transverse | |
| Infraorbital | |
| Spinosum | |
| Rotundum | |
| Ovale | |

## Exercise 5: Bony landmarks of the appendicular skeleton

In this exercise, you will review the bony landmarks found on the appendicular skeleton. For these bones, you will need to be able to identify left vs. right.

*Scapula (anterior, posterior, and later views):*

- ☐ Coracoid process
- ☐ Acromion
- ☐ Glenoid fossa/cavity
- ☐ Superior angle
- ☐ Inferior angle
- ☐ Lateral border
- ☐ Supraglenoid tubercle (barely shown)
- ☐ Supraspinous fossa
- ☐ Infraspinous fossa
- ☐ Spine
- ☐ Suprascapular notch

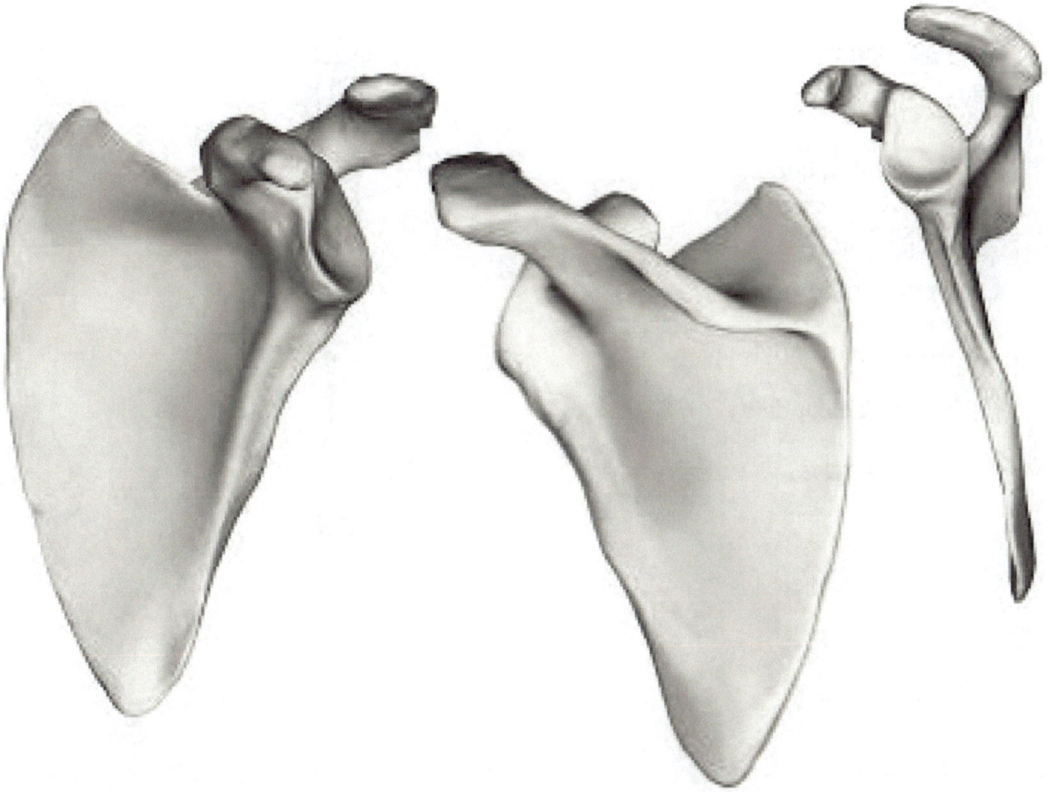

*Clavicle (superior view, left and right):*

- [ ] Sterni end
- [ ] Acromial end
- [ ] Conoid tubercle

*Humerus:*

- [ ] Greater tubercle
- [ ] Lesser tubercle
- [ ] Intertubercular sulcus
- [ ] Coronoid fossa (small)
- [ ] Radial fossa (small)
- [ ] Olecranon fossa
- [ ] Trochlea
- [ ] Capitulum
- [ ] Head
- [ ] Anatomical neck
- [ ] Surgical neck
- [ ] Deltoid tuberosity
- [ ] Medial epicondyle
- [ ] Lateral epicondyle
- [ ] Radial groove

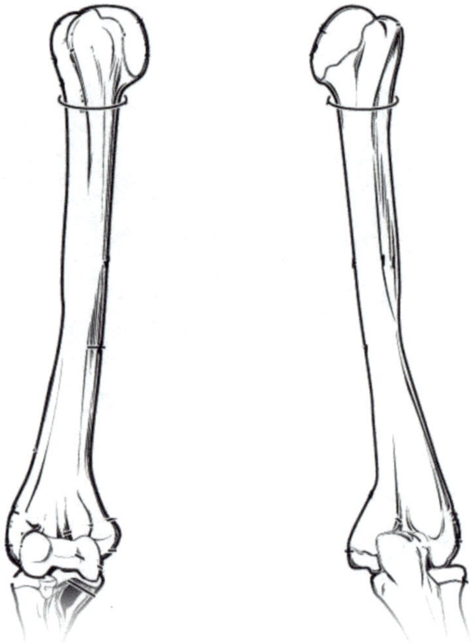


6

[13] "Clavicle" by BodyParts3D is licensed under CC BY-SA 2.1 Japan

[14] "Humerus, Radius, and Ulna" by OpenStax is licensed under CC BY 3.0.Deleted Labels.

*Radius and Ulna:*

- [ ] Olecranon
- [ ] Styloid processes (of radius and ulna)
- [ ] Head of radius
- [ ] Radial tuberosity
- [ ] Radial notch (of unla)
- [ ] Coronoid process
- [ ] Ulnar notch (of radius)

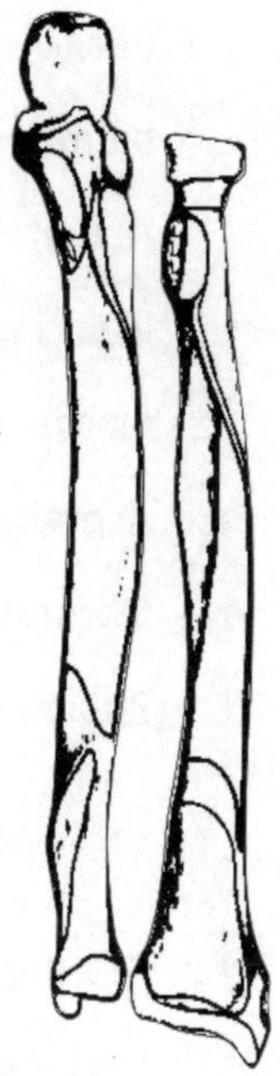

*Hand (left, anterior view):*

- [ ] Scaphoid
- [ ] Lunate
- [ ] Triquetrum
- [ ] Pisiform
- [ ] Hamate
- [ ] Capitate
- [ ] Trapezoid
- [ ] Trapezium
- [ ] Metacarpals (5x)
- [ ] Phalanges (14x)

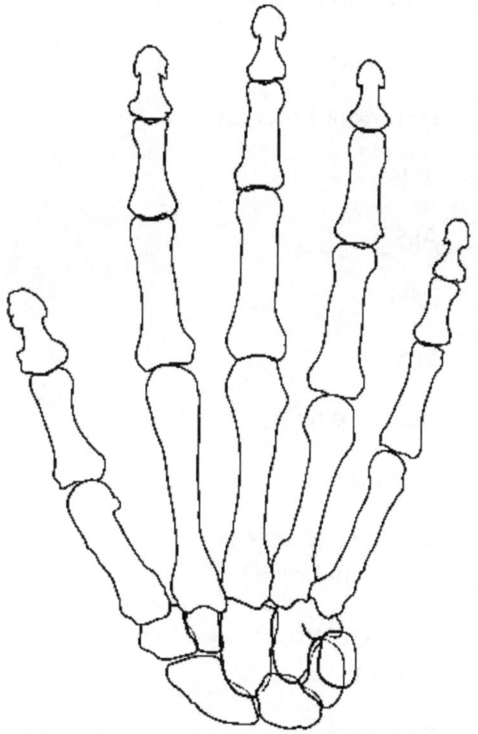

*Os coxae (incl. rest of pelvic girdle):*

- ☐ Pubis (outline with color)
- ☐ Ischium (outline with color)
- ☐ Pelvic wing (ala)
- ☐ Ilium (outline with color)
- ☐ Arcuate line
- ☐ Iliac crest
- ☐ Iliac fossa
- ☐ Acetabulum
- ☐ Obturator foramen
- ☐ Pubic symphysis (joint)

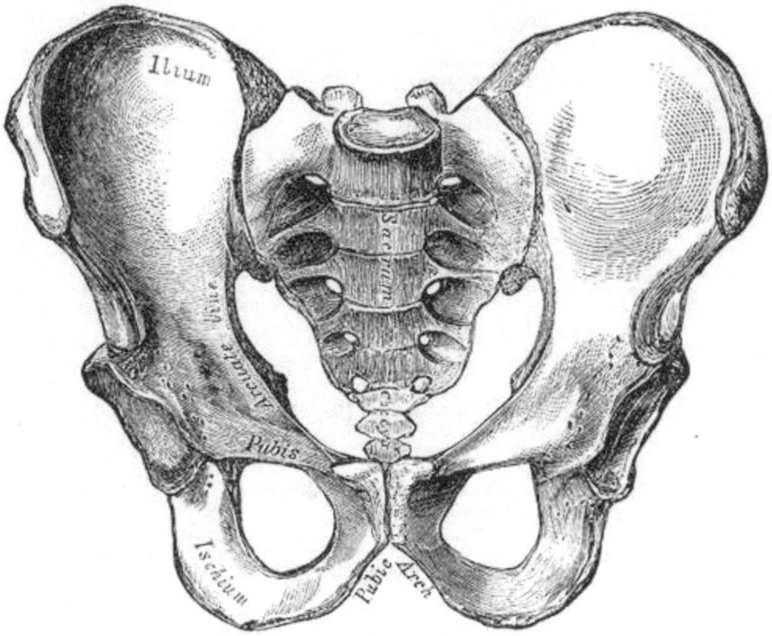


*Os coxae (lateral):*

- ☐ Greater sciatic notch
- ☐ Lesser sciatic notch
- ☐ Acetabulum
- ☐ Obturator foramen
- ☐ ASIS
- ☐ AIIS
- ☐ PSIS
- ☐ PIIS
- ☐ Ischial Spine
- ☐ Pubic tubercle
- ☐ Sacrum and coccyx (separate bones)
- ☐ Ischial tuberosity (not drawn)
- ☐ Iliac crest

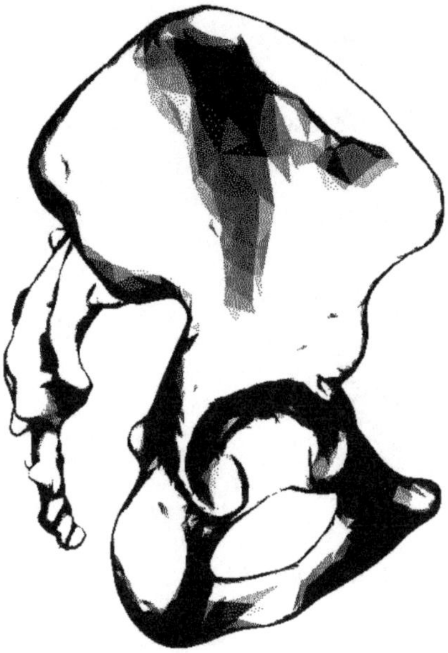

6

[15] BodyParts3D/Anatomography. Creative Commons Attribution 2.1 Japan.

*Femur (anterior and posterior view):*

- ☐ Greater trochanter
- ☐ Lesser trochanter
- ☐ Intertrochanteric line
- ☐ Intertrochanteric crest
- ☐ Fovea capitis
- ☐ Linea aspera and gluteal tuberosity
- ☐ Head
- ☐ Neck
- ☐ Condyles (medial and lateral)
- ☐ Intercondylar fossa
- ☐ Epicondyles (medial and lateral)

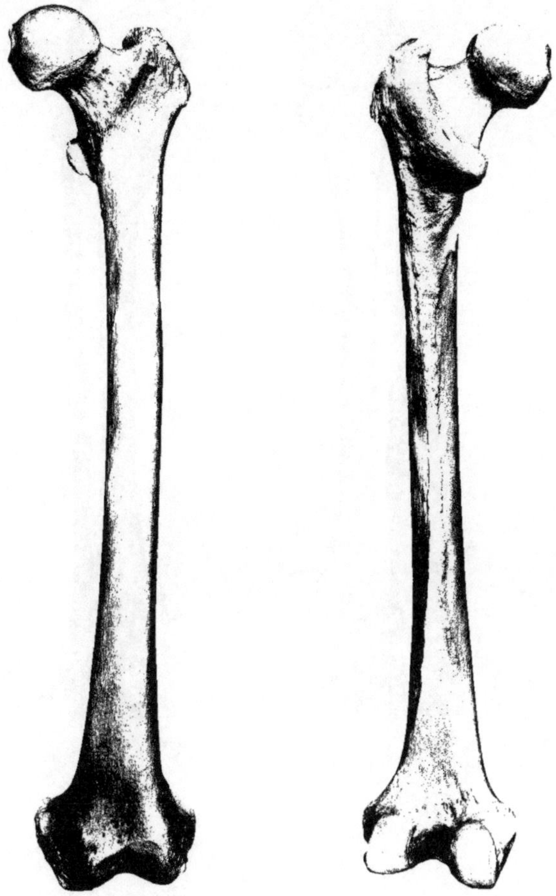

*Tibia and Fibula:*

- ☐ Tibial tuberosity (barely visible)
- ☐ Medial malleolus
- ☐ Lateral malleolus

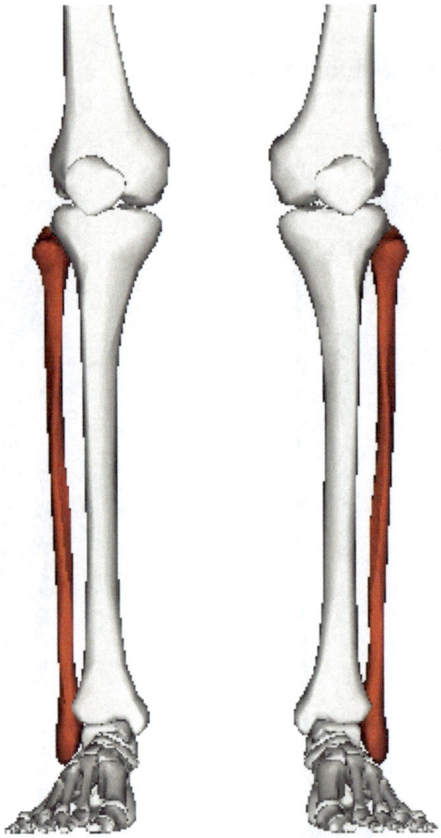

6

[16] "Tibia and Fibula" by BodyParts3D is licensed under CC BY-SA 2.1 Japan

*Patella:*

- ☐ Base
- ☐ Apex

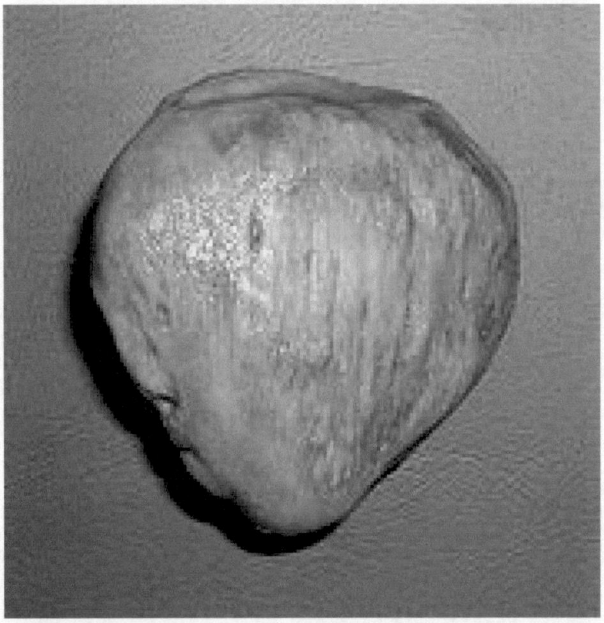

6

*Foot (anterior view):*

- ☐ Talus
- ☐ Calcaneus (barely visible)
- ☐ Cuboid
- ☐ Navicular
- ☐ Cuneiforms (M,I,L)
- ☐ Metatarsals
- ☐ Phalanges (14)

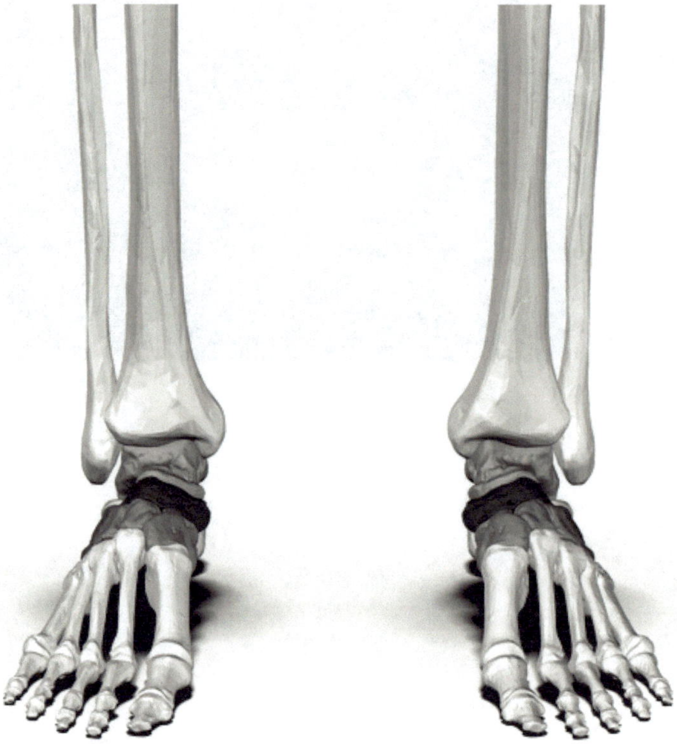

6

[17] "Tibia, Fibula, and Foot bones" by BodyParts3D is licensed under CC BY-SA 2.1 Japan

## Review questions:

1. Your first digit on your hand is your ________________ and your fifth digit is your ________________ finger.

2. Your first digit on your foot is your ________________ and your fifth digit is your ________________ toe.

3. All your fingers and toes have 3 phalanges besides for your ________________ and ________________.

6

Intentionally left blank

6

# Lab 7

# Joints

7

## Learning Objectives: Students will be able to…

- Describe the various types of joints
- Describe the various types of synovial joints, and their range of motion
- Identify structures of major joint systems in the human body

## Exercise 1: Basics of joints

Joints are formed by the articulation ("joining") of bones. Acting as glue are **ligaments** and other connective tissues. Joints are categorized into three major categories: **syndesmotic**, **amphiarthrotic**, and **diarthrotic** ("synovial joints"). Syndesmotic joints allow for no movement, amphiarthrotic allow for some, and diarthrotic allow for "free" movement. Each type of joint can then broken down into subcategories:

Table 1. Types of Joints

| Joint Type | Subcategory | Example |
|---|---|---|
| **Syndesmotic** | Suture | Coronal suture of the skull |
| | Gomphosis | Teeth to the mandible |
| | Synchondrosis | Costal cartilage linking ribs to sternum |
| | Synostosis | Fusion of illium, ischium and pubis of *os coxae* |
| **Amphiarthrotic** | Symphysis | Pubic symphysis linking together the two *os coxae* |
| | Syndesmosis | Crural and antebrachial interosseus membranes |
| **Diarthrotic** | See below | |

Synovial joints are more complex, and they are described based on the movements they allow. As you will see, they don't truly allow for free movement. Compared to other types of joints though, they generally have a much greater range of motion. Synovial joints also have unique structure. They are known for having a capsule lined with a synovial membrane. Within this capsule is a synovial fluid which lubricates the joint and allows for force transfer between bones. The image on the right shows a typical synovial joint. The image below demonstrates the six types of synovial joints

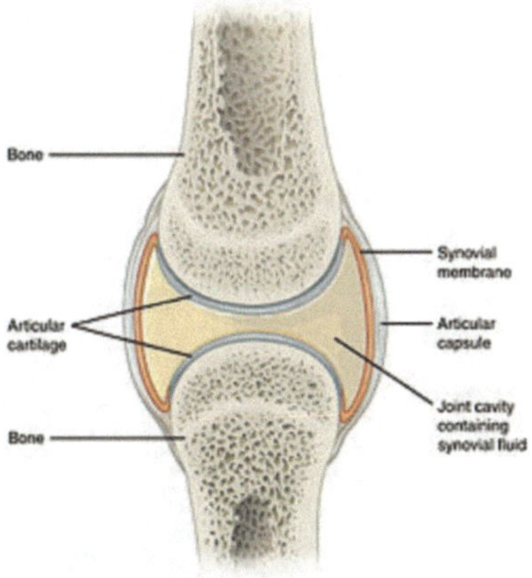


[1] "Synovial Joint" by OpenStax is licensed under CC BY 4.0 International.

## Clinical Application: Women and Ligament Tears

The pelvic girdle of women has evolved to accommodate child birth. Due to our very large brains and skulls, human child birth is one of the most dangerous in the animal kingdom. Unfortunately, the wider pelvic girdle means that the femur is angled towards the midline of the body. In men, the femur is more vertical. This is known as the **Q angle**. Women have a larger Q angle then men, and this gives rise to issues for the knee joint.

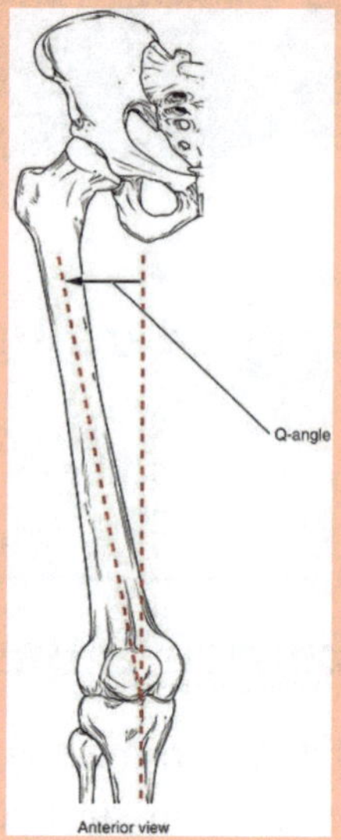


In particular, this puts enormous stress on the **anterior cruciate ligament**. You might know this ligament by its abbreviation, the ACL. The ACL prevents forward movement (**anterior translation**) of the tibia. Considering evolution once again, the ACL did not strengthen as humans evolved to walk upright. Compared to the quadrupeds, bipedal walking strains the ACL, and ours cannot handle this stress. It is no surprise then that sports like soccer, which require kicking and running, can cause ACL tears. Combined with the natural strain on the ACL for women, women soccer players must be especially careful.

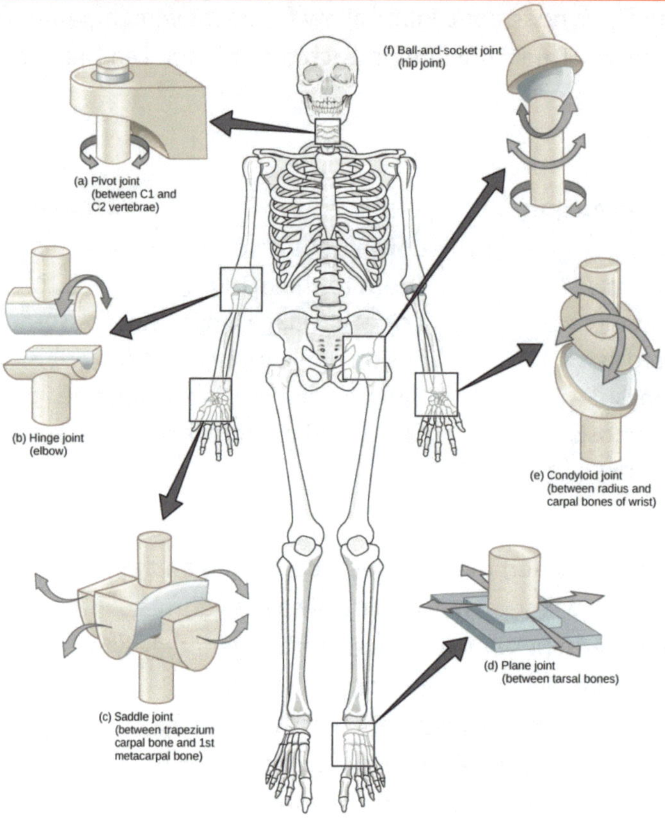


[3]

[2] "Types of Synovial Joints" by OpenStax is licensed under CC BY 4.0 International.

[3] "Q angle of knee" by Abdallah Ahmed is licensed under CC BY-SA 3.0 Unported

## Review questions:

1. _________________ joints allow for no movement
2. _________________ joints allow for some movement
3. _________________ joints for free movement
4. Your tooth-mandible joint is described best as a _________________ joint
5. Your os coxae-os coxae joint is described best as a _________________ joint

7

## Application questions:

1. Reviewing the image above, which two synovial joints allow for the least amount of movement?

2. Reviewing the image above, what synovial joint allows for the greatest amount of movement?

3. What type of joint would your knee be?

4. The synovial membrane is composed of what type of tissue?

# Exercise 2: Major synovial joint systems

As you work through these next pages you will be coloring (and optionally numbering) each ligament and structure of the following joint systems. Color each box next to the image and use the same color to color in the drawing. Add numbers if you wish. Do not label the complete name directly on diagram, so that you can study from these pages later without seeing the answers.

## Elbow Joint (lateral):

- ☐ Ulnar collateral ligament
- ☐ Radial collateral ligament
- ☐
- ☐ Articular capsule
- ☐ Annular ligament

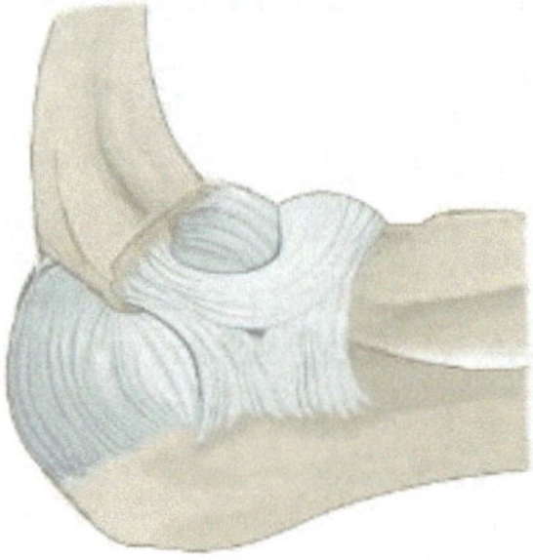

(b) Lateral view of right elbow joint

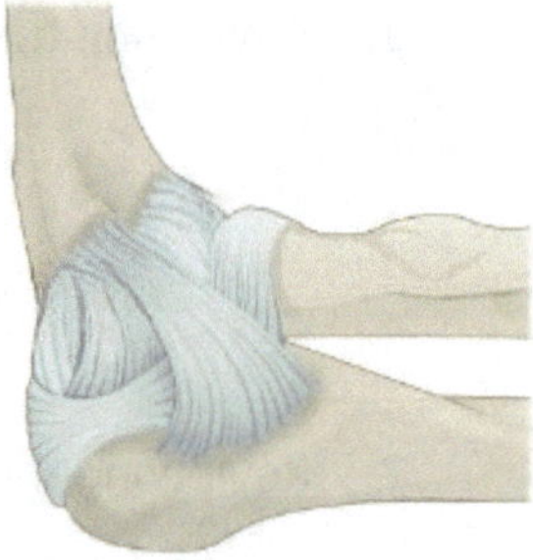

(c) Medial view of right elbow joint

[5]

## Knee Joint:

- ☐ Anterior cruciate ligament
- ☐ Posterior cruciate ligament
- ☐ Medial collateral ligament (not shown)
- ☐ Lateral collateral ligament (not shown)
- ☐ Meniscus (medial and lateral)

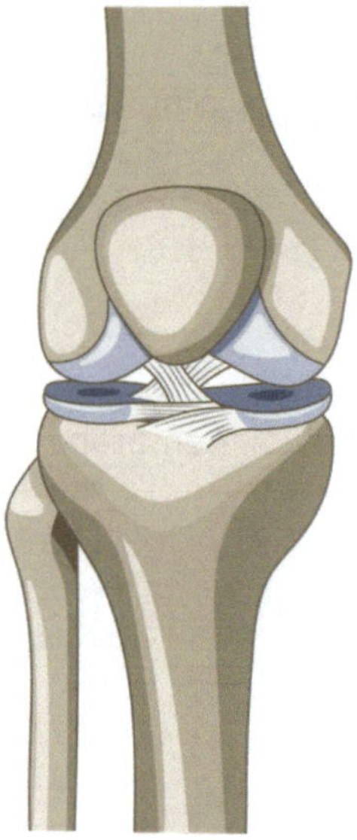

---

## Shoulder Joint:

- [ ] Coracoclavicular ligament (2)
- [ ] Coracoacromial ligament
- [ ] Acromioclavicular ligament
- [ ] Bursa (for deltoid muscle, not shown)
- [ ] Articular capsule

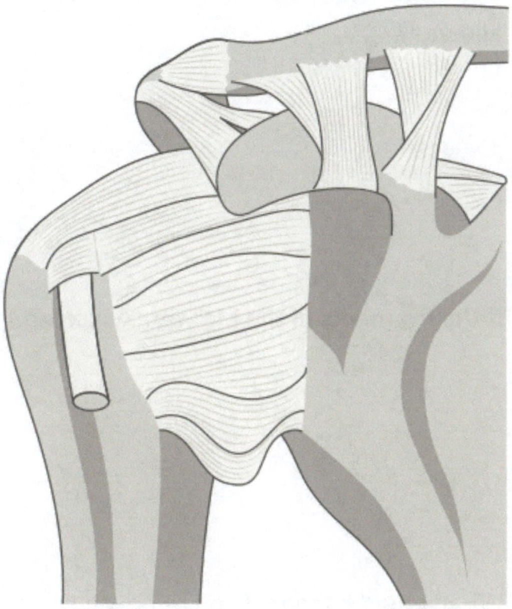

## Hip Joint:

- [ ] Illiofemoral ligament
- [ ] Ischiofemoral ligament
- [ ] Pubofemoral ligament

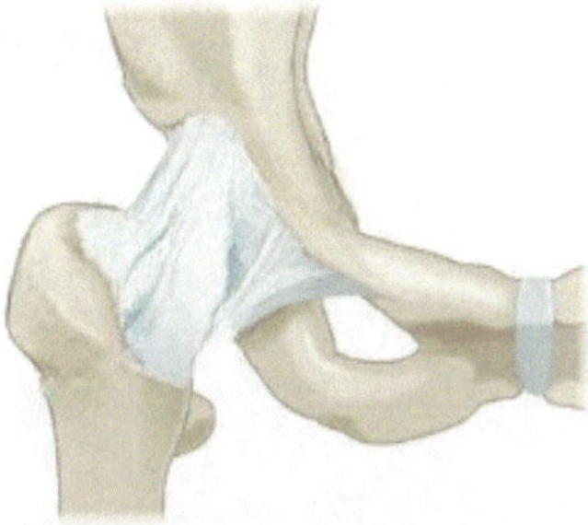

(b) Anterior view of right hip joint, capsule in place

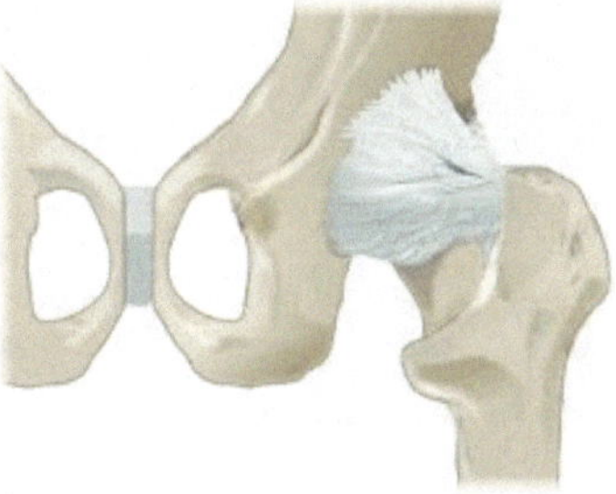

(c) Posterior view of right hip joint, capsule in place

[6] "Hip Joint" by Openstax is licensed under CC BY 4.0

[7] " Ligament of the shoulder joint" by DataBase Center for Life Science is licensed under CC BY 4.0

7

## Application questions:

1. If the posterior cruciate ligament of the knee was cut, what motion of the knee would be allowed?

2. If the ligaments of the hip were weakened, what motion of the hip would be allowed?

3. What would happen if you removed the bursa of the shoulder joint?

# Lab 8

# Muscles

## Learning Objectives: Students will be able to…

- Describe the overall structure of muscle tissue
- Compare and contrast skeletal, cardiac and smooth muscle structurally and functionally, gross and microscopic
- Identify and name certain facts about the muscles listed in your lab manual on anatomical models and pictures
- Identify major muscles of the human body, and list key facts about them (e.g., innervation, insertion, etc.)

8

## Exercise 1: Muscle histology

Muscles are how your body moves. By contracting they can act upon joints to change the angle of articulating bones causing movement of the skeleton. The connection points for muscles, the **origins** and **insertions**, are usually bony landmarks. Origins are furthest away from the join and insertions are closet to the joint. The above characteristics are particularly true for skeletal muscle.

There are 3 types of muscle tissue: **skeletal**, **smooth** and **cardiac**. When people are talking about muscles in normal conversation, they are usually discussing skeletal muscles. These are the muscles that are important for movement and stabilization of the body. Smooth muscle and cardiac are important too, however. The former lines many visceral organs such as the intestines. The later makes up the contractile tissue of our heart. Muscle also forms the 3$^{rd}$ of the 4 major tissue types.

As you have seen, the human body is connected. The bones, joints, and muscles work together to allow movement. It is no surprise then, that the final type of tissue, nervous tissue, is also connected to muscular contraction. Driving all of these types of muscles is the nervous system. Your brain, brain stem, and spinal cord control muscular contraction.

You will first draw images of histological slides of muscle tissue in order to better understand how these tissues are structured.

## Skeletal muscle (20)

With its striated pattern, skeletal muscle is easily identifiable. The fibers you will see will be continuous with no breaks in your field of view. Depending on what muscle was used to make the slide, the tissue will be a darker pink color. There also will be plenty of nuclei present. Skeletal muscle tissue as it ages tend to accumulate nuclei

**Label the following**: Myocyte/"muscle fiber", nuclei



Total Magnification:

## Cardiac muscle (21)

Cardiac muscle is very similar in appearance to skeletal muscle. You will see similar striations and staining as with skeletal muscle. One major difference is the presence of intercalated discs. The shape, overall, of cardiac muscle is slightly different as well. You may notice more branching in cardiac muscle.

**Label the following**: Cardiomyocyte, nuclei, intercalated disc



Total Magnification:

## Smooth muscle (22)

Without distinct sarcomeres, smooth muscle lacks striation. In some regards, smooth muscle looks similar to dense regular tissue, but the nuclei are not spindle shaped and are round instead. Longer spindle shaped fibers and myocytes may be present as well.

**Label the following**: Myocyte, nuclei



Total Magnification:

## Review questions:

1. Fill out this table comparing the three muscle tissue types

| Tissue | Striations? | Intercalated discs? | Nuclei? |
|---|---|---|---|
| **Skeletal muscle** | | | |
| **Smooth muscle** | | | |
| **Cardiac muscle** | | | |

## Application questions:

1. Even though the intercalated discs appear as "walls" separating the cardiomyocytes they are actually gap junctions. What do you think is passing through these gap junctions? Why would this be in particular very important for the heart?

8

## Exercise 2: Myocyte structures

**Myocytes** ("muscle cells") are quite different from what one likely thinks of when imagining a cell. Myocytes often have multiple nuclei and mitochondria; both of which are required for the high-energy consumption of the myocyte. There is also a modified endoplasmic reticulum called the **sarcoplasmic reticulum**. This is specialized for storage of calcium. Another unique structure of the myocyte are invaginations of the plasma membrane called **transverse tubules** (T tubules). Muscular contraction is driven by electrochemical gradients that surround the myocyte, so the transverse tubules allow that electrochemical gradient to permeate inwards. Straddling the transverse tubule are **terminal cisternae**, enlargements of the sarcoplasmic reticulum. Together, 1 transverse tubule and 2 terminal cisternae make a **triad**. Despite all of this the actual function unit of the muscle, the part that contracts, is the **sarcomere**. Sarcomeres line up to form **myofibrils**. Zooming in, sarcomeres are composed mainly of two proteins - **actin** and **myosin**. These aggregate into larger structures called the **thin** and **thick** filaments, respectively.

On the image below, label the following structures.

- Mitochondria
- Sarcomere and myofibrils
- T-tubules
- Terminal Cisternae of Sarcoplasmic reticulum (SR)
- Triad
- Sarcolemma

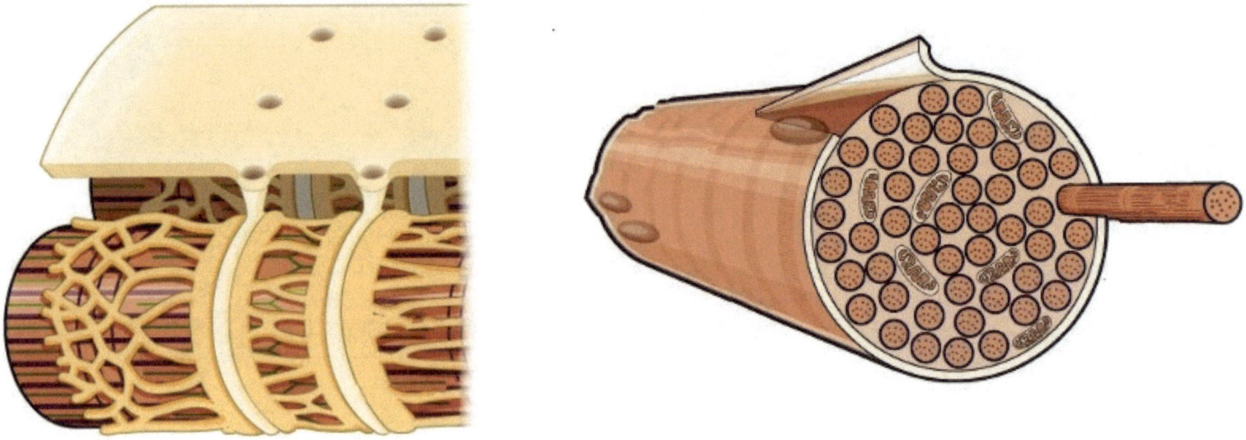

[1] "Myocyte" by OpenStax is licensed under CC BY 4.0

## Review questions:

1. A triad consists of what three parts of the myocyte?

2. The "functional unit" of muscle is the....



## Application questions:

1. Why does skeletal muscle have a particularly high number of mitochondria?

2. Skeletal muscle contains a protein called myoglobin that captures oxygen just like hemoglobin in blood. What is it important for muscles to get oxygen?

## Exercise 3: Major muscles of the human body

The number of bones in the human body is paltry compared to the number of muscles. The number varies, but there are approximately 650 named muscles. Part of the variability is a result of the great variation there is in human musculature. Muscles like the psoas minor are thought to be missing in about 50% of the population.

### Every-body is Different: The Palmaris Longus Muscle

Our musculature is far from uniform. There are numerous, tiny muscles of the human body that may be absent. One great example of this is the palmaris longus. The absence of the palmaris longus has no appreciable difference in grip or pinch strength in adults. Interestingly, the prevalence of the palmaris longus varies in different populations. For example, 22.6% of Turkish people are missing the palmaris longus, whereas only 4.8 % of Asian and 3.0 % of Black populations are.

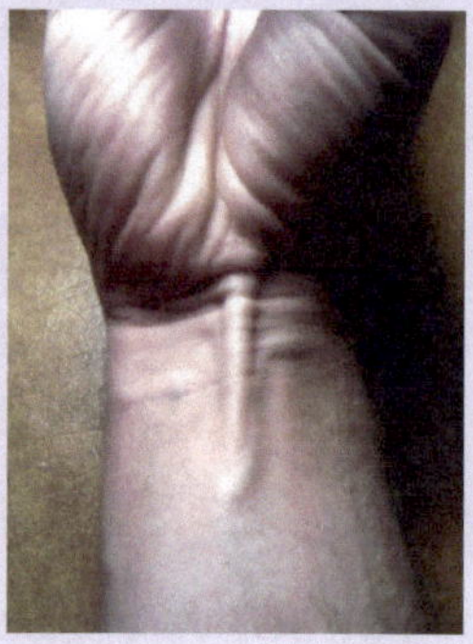

You can easily test to see if you have a palmaris longus by doing Schaeffer's test. Simply bring your oppose your first and fifth finger, and the flex your wrist at the same time. If you have the palmaris longus, you will see a bulging tendon.

8

While there are many muscles to learn, a selection are included in the following section so that they can be labeled.

As you work through these next pages you will be coloring (and optionally numbering) each muscle. Color each box to the left of the diagram and use the same color to color in the diagram. Add numbers if you wish. Do not label the complete name directly on the diagram, so that you can study from these pages later without seeing the answers.

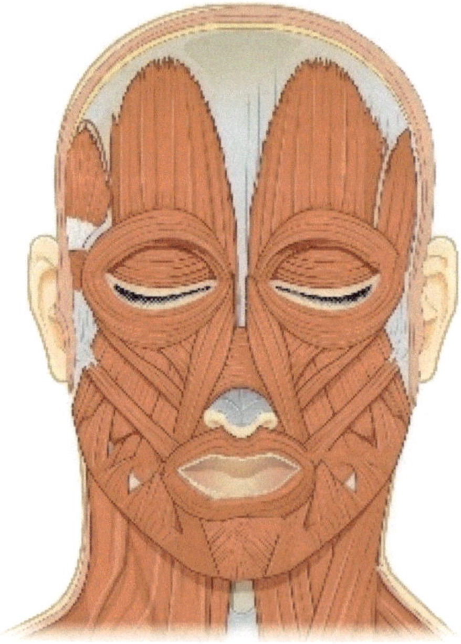

### Muscles of the face:

- ☐ Frontalis
- ☐ Orbicularis oculi
- ☐ Zygomaticus major
- ☐ Zygomaticus minor
- ☐ Temporalis
- ☐ Orbicularis oris
- ☐

- [ ] Depressor anguli oris
- [ ] Nasalis
- [ ] Galea aponeurosis

## Muscles of the lateral skull:

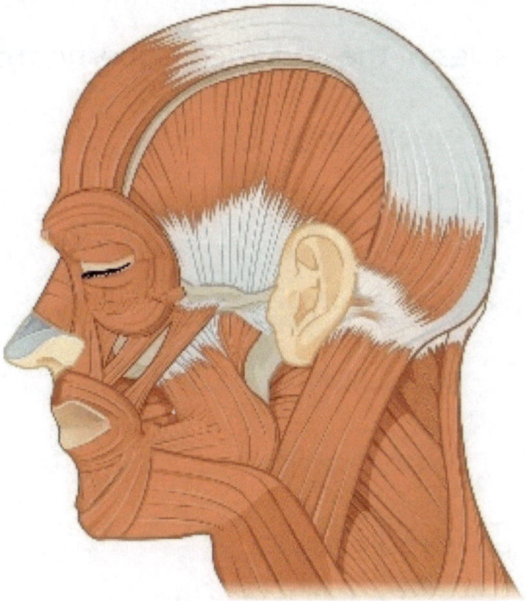

- [ ] Temporalis
- [ ] Occipitalis
- [ ] Sternocleidomastoid
- [ ] Platysma
- [ ] Trapezius
- [ ] Buccinator
- [ ] Masseter

## Muscles of the anterior neck:

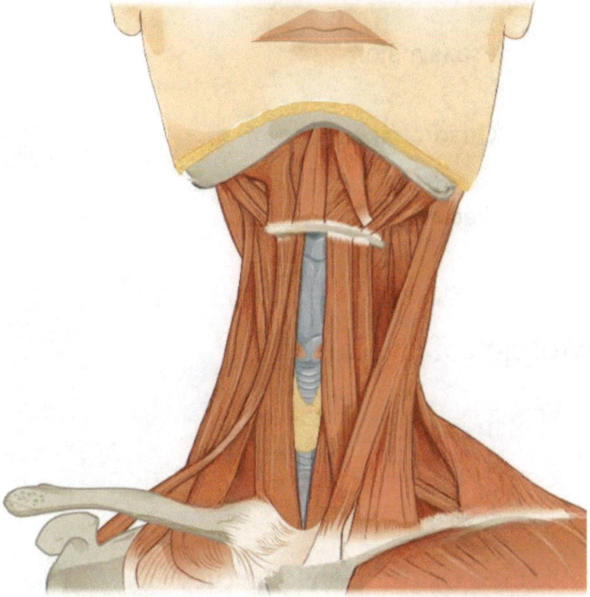

- [ ] Geniohyoid
- [ ] Mylohyoid
- [ ] Thyrohyoid
- [ ] Omohyoid
- [ ] Sternohyoid

## Muscles of the thorax:

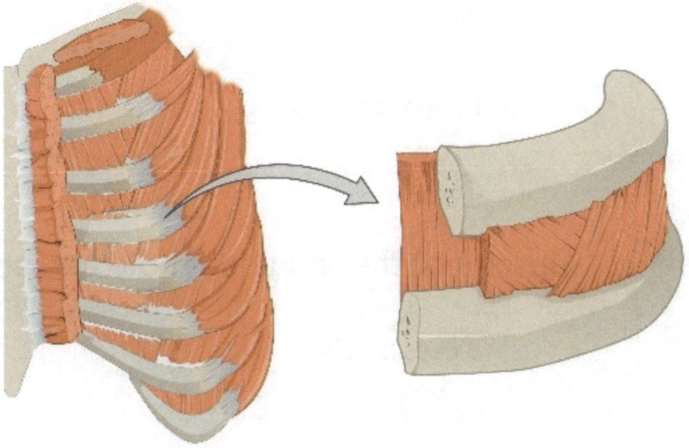

- [ ] Pectoralis minor
- [ ] Serratus anterior
- [ ] External intercostals
- [ ] Internal intercostals

- ☐ Innermost intercostals

## Muscles of the abdomen (superficial):

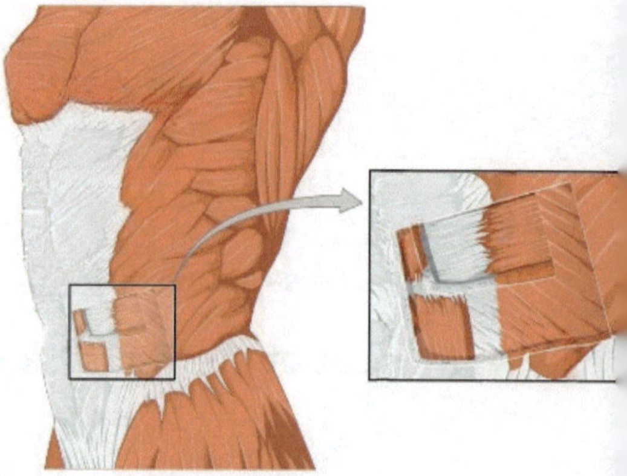

(a) Superficial and deep abdominal muscles (anterior lateral view)

- ☐ Pectoralis major
- ☐ Serratus anterior
- ☐ External obliques
- ☐ Rectus sheath
- ☐ Rectus abdominis
- ☐ Internal oblique

## Muscles of the abdomen (deep):

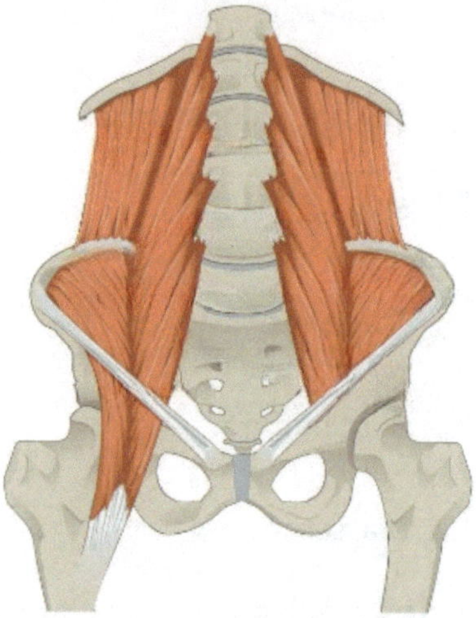

(b) Posterior abdominal muscles (anterior view)

- ☐ Psoas major
- ☐ Quadratus lumborum
- ☐ Iliacus

## Muscles of the back:

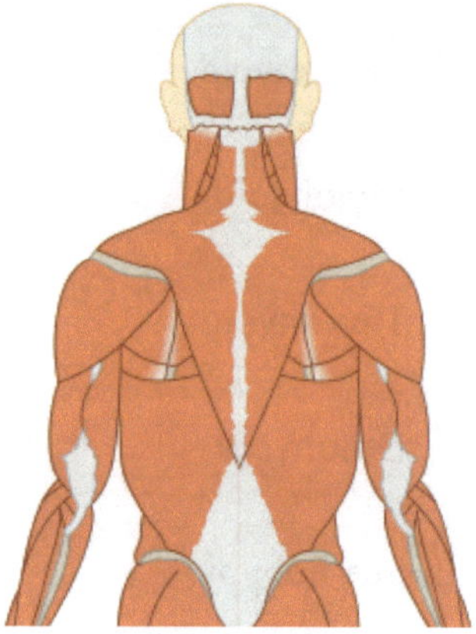

- ☐ Trapezius
- ☐ Latissimus dorsi

## Muscles of the pelvic girdle:

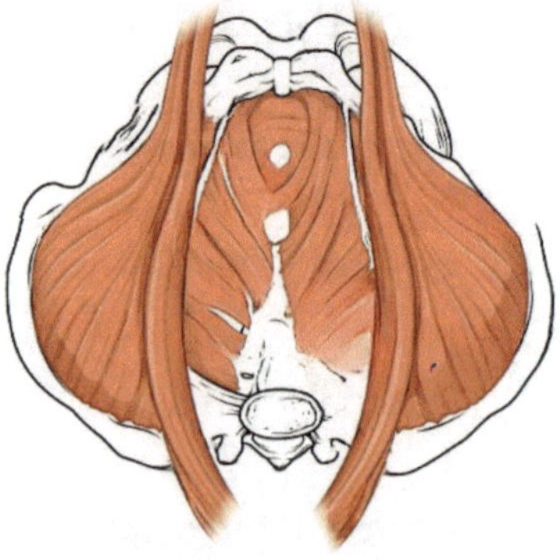

- ☐ Iliococcygeus
- ☐ Pubococcygeus
- ☐ Iliacus

## Muscles of the shoulder girdle and arm:

- ☐ Biceps brachii

- ☐ Brachialis
- ☐ Triceps brachii
- ☐ Supraspinatus
- ☐ Infraspinatus
- ☐ Teres major
- ☐ Teres minor

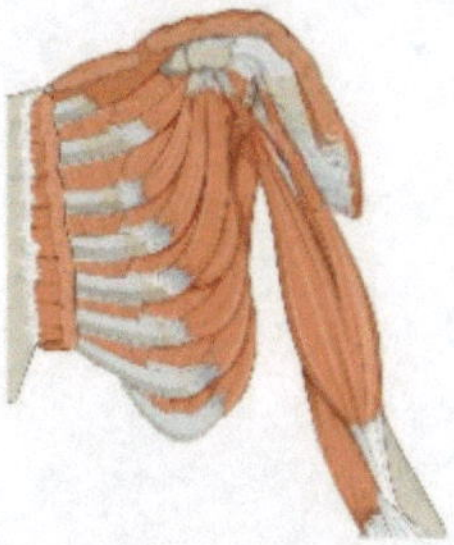

Left upper arm muscles (anterior lateral view)

Left upper arm muscles (posterior view)

## Muscles of the forearm:

- ☐ Brachioradialis
- ☐ Pronator teres
- ☐ Flexor carpi ulnaris
- ☐ Extensor digit minimi
- ☐ Extensor digitorum
- ☐ Extensor carpi ulnaris

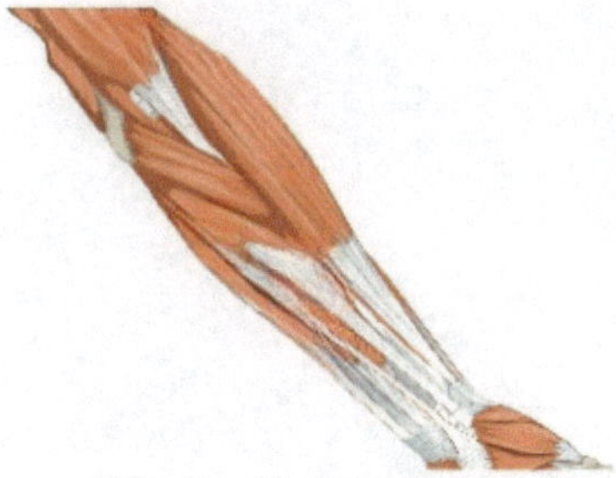

Left forearm superficial muscles (palmar view)

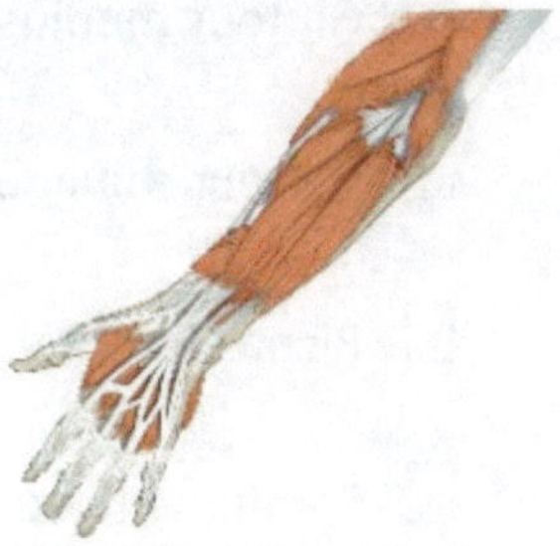

Left forearm superficial muscles (dorsal view)

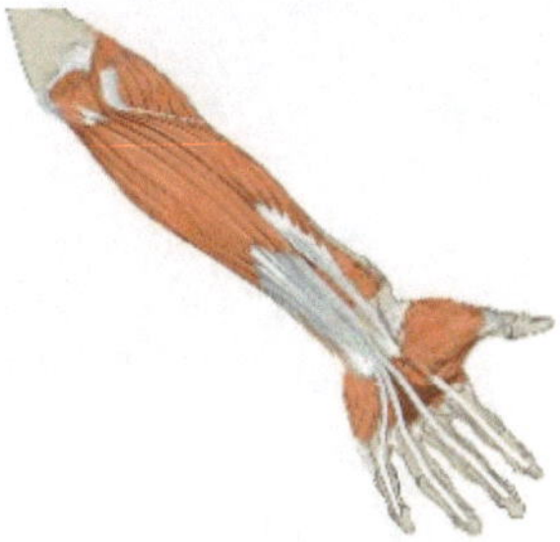

Left forearm deep muscles (palmar view)

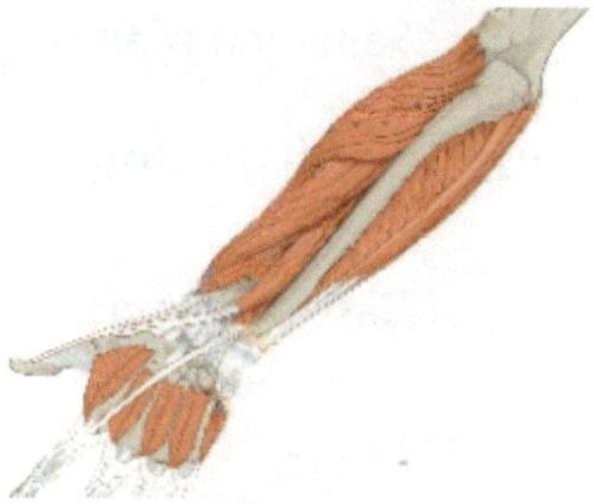

Left forearm deep muscles (dorsal view)

## Muscles of the thigh (anterior):

- ☐ Sartorius
- ☐ Rectus femoris
- ☐ Vastus lateralis
- ☐ Vastus medialis
- ☐ Gracilis
- ☐ Adductor magnus
- ☐ Adductor brevis

8

- [ ]
- [ ] Adductor longus

## Muscles of the thigh (posterior):

- [ ] Gluteus maximus (cut)
- [ ] Gluteus medius (cut)
- [ ] Gluteus minimus
- [ ] Biceps femoris
- [ ] Semitendinosus
- [ ] Semimembranosus

## Muscles of the leg:

- [ ] Gastrocnemius
- [ ] Soleus [3]
- [ ] Tibialis anterior

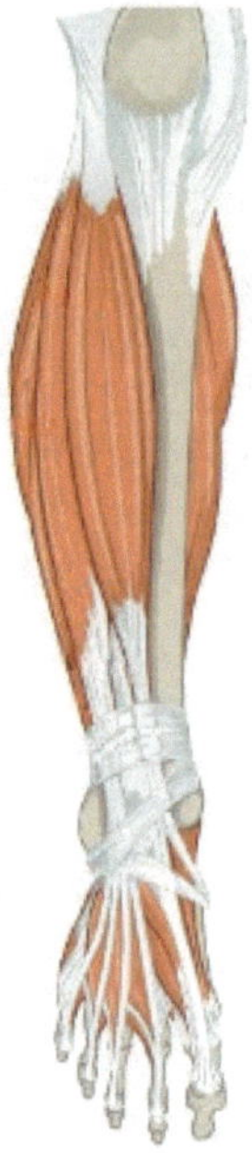

[2] "Muscles of the Face", "Muscles of the Anterior Neck", "Muscles of the Thorax", "Muscles of the Abdomen", "Muscles of the Thigh", and "Muscles of the Leg", "Muscles of the Pubic Floor", "Muscles of the Upper Limb" by Openstax is licensed under CC BY 4.0

[3] "File:1105_Anterior_and_Posterior_Views_of_Muscles.jpg" by Tomáš Kebert is licensed under CC BY-SA 4.0 / A derivative from the original work

## Review questions:

1. Muscles are best learned as parts of larger groups. Below are some major muscle groups, identify all members for each group.

    a. "Quadriceps"

    b. "Hamstrings"

    c. Extrinsic eye muscles

    d. Erector spinae

    e. Lateral rotators of the hip

    f. Adductors of the hip

    g. "Rotator cuff" muscles

8

2. One way to conceptualize muscles is think of agonist-antagonist pairs. Agonists perform the action, and antagonists perform the opposite action. Identify the antagonist for each muscle or group below.

   a. Rectus abdominis:______________________________

   b. Biceps brachii:______________________________

   c. Quadriceps:______________________________

   d. Gastrocnemius and soleus:______________________________

# Lab 9

# The Nervous System

## Learning Objectives: Students will be able to...

- Give an overview of the Central (CNS) and Peripheral (PNS) Nervous Systems
- When given a diagram or model of a neuron, identify key parts
- When given a diagram of nerve, identify key parts
- List major neuroglia and their functions

## Exercise 1: Overview of the nervous system

The nervous system is a complex arrangement of nervous (and other) where the major function is to relay signals called **action potentials**. These electrical impulses travel at about a rate of ~200 MPH! This is critical because the nervous system is about responding to quick reactions to the environment. If you are trying to escape a tiger or wolf you would want the nervous system to drive contractions of your skeletal muscle to make sure you escaped quickly. Similarly, you want your sensory organs (e.g. eyes and ears) to be linked to the nervous system to ensure that you are actually aware there is even a tiger there quickly enough.

The major cell of nervous tissue is the **neuron**. These are the cells that conduct action potentials. On top of that, there exists a handful of **neuroglia**. These are support cells of nervous tissue. For example, the microglia are responsible for cleaning up debris and pathogens; they function like immune cells, but are specific to the nervous system. Neurons aggregate into larger structures. The brain itself is a meshwork of neurons and neuroglia. If we are thinking about peripheral structures of the body such as your upper limb, nervous tissue aggregates into **nerve fibers**. This allows for neurons to extend long distances such as from your spinal cord to the tips of your fingers. Part of this ability to extend long distances is due to the **axon**. This is a long extension of the cytoplasm and plasma membrane

There are many divisions of the nervous system, but the two major ones are the **central** and **peripheral nervous systems**. The central nervous system (CNS) is composed of the brain, cerebellum, brainstem, and spinal cord. The peripheral nervous system (PNS) is then composed of any nerve fibers extending off of those structures.

The below questions summarize some salient points of our lecture, and review major concepts of the nervous system.

## Review questions:

1. What is the "opposite" division of the nervous system for term below?
   a. CNS:____________________
   b. Sensory: ____________________
   c. Somatic nervous system:____________________
   d. Sympathetic nervous system:____________________
2. The dominant (and perhaps most important) cell of the nervous system is the

   ____________________.
3. What is the difference between a nerve and a neuron?
4. The "on switch" for neuron is called an ____________________.
5. Neurons form two major types of synapses between themselves. What are they? Describe each.
6. The synapse between a neuron and a muscle is called the ____________________.
7. What is grey matter? What is white matter?

9

8. There are various neuroglia that make up the nervous system. Fill out the table below that summarizes their major features.

| Neuroglia | CNS or PNS? | Function |
|---|---|---|
| Astrocytes | | |
| Microglia | | |
| Ependymal cells | | |
| Schwann Cells | | |
| Satellite Cells | | |
| Oligodendrocytes | | |

9

## Application questions:

1. If an electrical impulse in a nerve travels at 200 MPH, how long would it take a signal from your brain to reach your toes? Assume the distance of this is 5 feet. Express the time of your final answer in seconds. (If you can't do this math, follow the footnotes below, but try your best to do it on your own!)

---

1. Convert miles to feet. 1 mile is 5280 feet. How many feet is 200 miles? You now have a number with units in feet/hour
2. Your number is in feet/hour. There are 3600 seconds in an hour, so your number from step 1 could also be written as # of feet/3600 seconds. If you divide your number by 3600 you will have a number in feet/second
3. Using your number from step 2, how many seconds would it take to travel 5 feet? Set up a proportion to solve this This number should be less than 1 second.
4. Your proportion will look like this: $\frac{5\ feet}{x\ second} = \frac{\#\ from\ step\ 2}{1\ second}$. Solve for x.

2. Using the appendix in your lab manual and/or the internet, determine the literal meaning of the following terms:

   a. Oligodendrocytes:____________________________

   b. Astrocyte:____________________________

   c. **Syn**apse:____________________________

   d. Retrograde:____________________________

   e. Saltatory:____________________________

   f. Ramus:____________________________

9

## Exercise 2: Neuron structure

The neuron is the main communicating cell of the nervous system. For the image below, label the 7 parts of the neuron below:

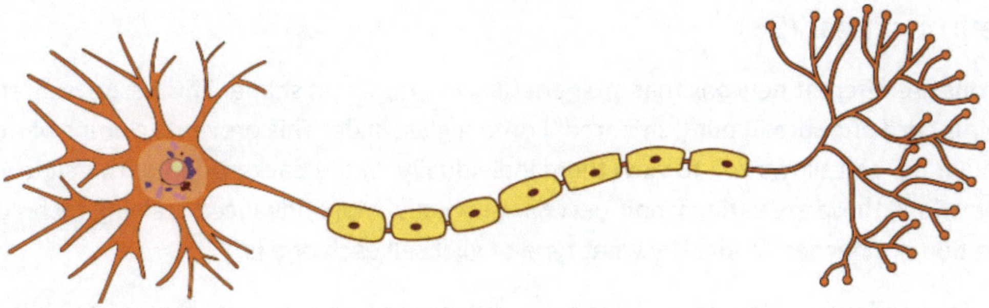

### Review questions:

1. Not shown here are Nissl bodies, what is the function of these structures?

2. To the neuron above, add an arrow(s) showing the flow of information. Where is information/neurotransmitters received and where is information/neurotransmitters released?

3. Circle an axon terminal in the image above. What is the difference between a chemical and electrical synapse in terms of what is happening at the axon terminal?

4. Label the axon hillock. What is special about the axon hillock?

### Application questions:

1. Is this neuron part of the CNS or PNS? How do you know? (hint: what cell is myelinating the axon?)
2. Discussed in class were three types of neurons: unipolar, bipolar, and multipolar. Which type is the neuron above?
3. Add a post-synaptic, multipolar neuron to the diagram above. Label the dendrites and axon of the neuron you have drawn.



[2] "Structure of Neuron (unlabeled)" by Sanu N is licensed under CC BY-SA 4.0

## Exercise 3: Nervous system histology

From the previous exercises, you reviewed major cells that compose the nervous system. In this exercise, you will explore nervous system histology

### Motor neuron smear (23)

Motor neurons are efferent neurons that are generally multipolar in shape. This tissue is often extracted from the spinal cord and spread out ("smeared") onto a glass slide. This prevents clumps of neurons from accumulating, and allows you to view them individually. In the background of this slide you will also see many dots, these are various, non-descript glial cells. More advanced staining techniques and slide preparation are needed to identify what type of glial cell each one is.

**Label the following**: Neuron, cell body, dendrite/axon (not easy to differentiate), glia

Total Magnification:

## Nerve, artery and vein cross-section (24)

Just like muscle fibers/myocytes, neurons are not individual cells floating around in your body. They are instead organized into structures called nerves. Similar to a muscle, each layer/level of the nerve has connective tissue surrounding it. The outermost later is epineurium, surrounding the fascicles (bundle of neurons) is the perineurium, and around each individual neuron is the endoneurium. Embedded with the nerve are then various blood vessels. This helps provide the neuron with nutrients and removes waste. Often these vessels will show as hollow circles on the slide.

**Label the following**: Epineurium, perineurium, nerve, fascicle, blood vessel

Total Magnification:

This page intentionally left blank

9

# Lab 10

# The Central and Peripheral Nervous Systems



## Learning Objectives: Students will be able to...

- Describe the gross anatomy of the length of spinal cord, naming the regions and structures.
- Identify any of the plexi and major spinal nerves of the human body
- Describe the cross-sectional anatomy of the "H" or "Butterfly" arrangement of the spinal cord.
- Give a description of the role and arrangement of the spinal meninges, the layers and the spaces.
- Identify major structures of the brain either on a diagram, human brain model, or preserved sheep brain
- Briefly explain the functions of each region and part of the brain (for example: ventricles, medulla, frontal lobe, etc.)
- Describe how CSF is circulated and its origin and function
- List and identify the 12 cranial nerves, and describe their functions

## Exercise 1: Spinal cord overview

The spinal cord as a whole has some major features that you should be aware of. For the picture to the right, circle and label the structures in the list below.

- Cauda equina
- Conus medullaris
- Posterior cervical enlargement (not shown)
- Lumbosacral enlargement (not shown)

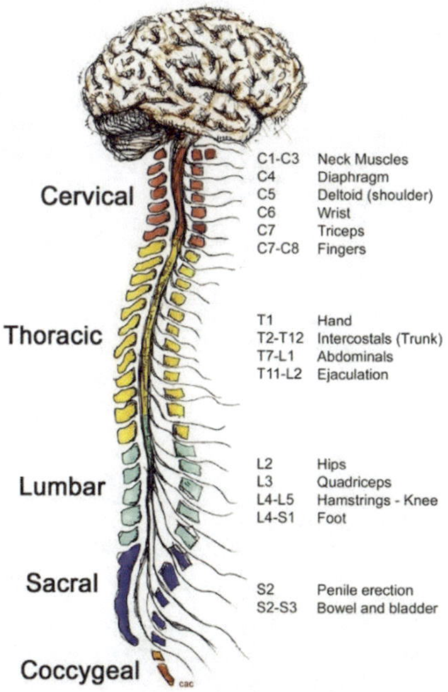


### Review questions:

1. How many cervical spinal nerves do you have?

[1] "Diagram showing the function of the nerves exiting each level of the spinal cord" by Vankadara Bhavya sree 1840585 is licensed under CC BY-SA 4.0

## **Exercise 2:** Spinal Cord Cross Section

A major aspect of the spinal cord is brought to light when viewing a region in cross-section. Below are two different representations of the spinal cord, one separated from the vertebral column, and one *in situ*. For each, label the spinal cords below. Note: not all structures may be found on each.

- Spinous process
- Body of vertebrae
- Transverse foramen
- Dorsal root
- Ventral root
- Ramus communicans (to be discussed later)
- Posterior median sulcus
- Anterior median fissure
- Rootlets
- A "nerve"
- Ventral ramus
- Dorsal ramus
- White columns
- Grey horns (anterior and posterior)
- Grey commissure
- Central canal (not shown)

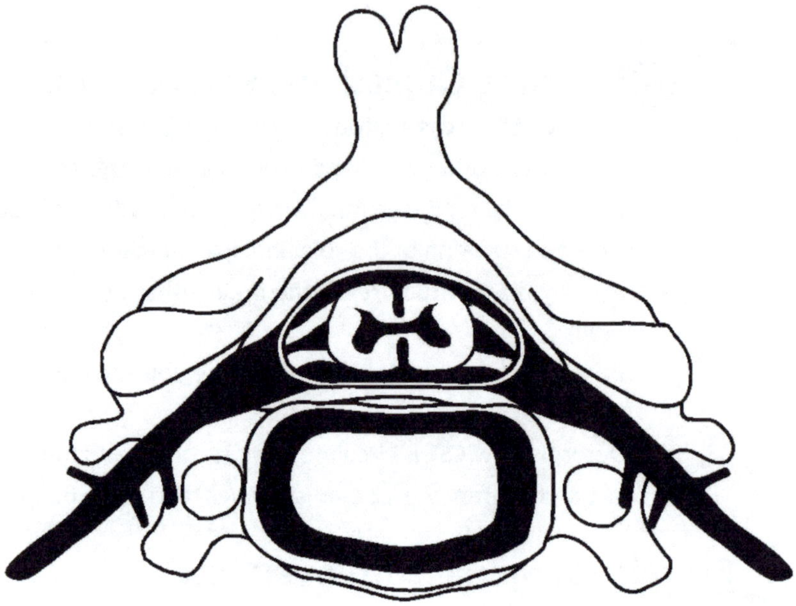



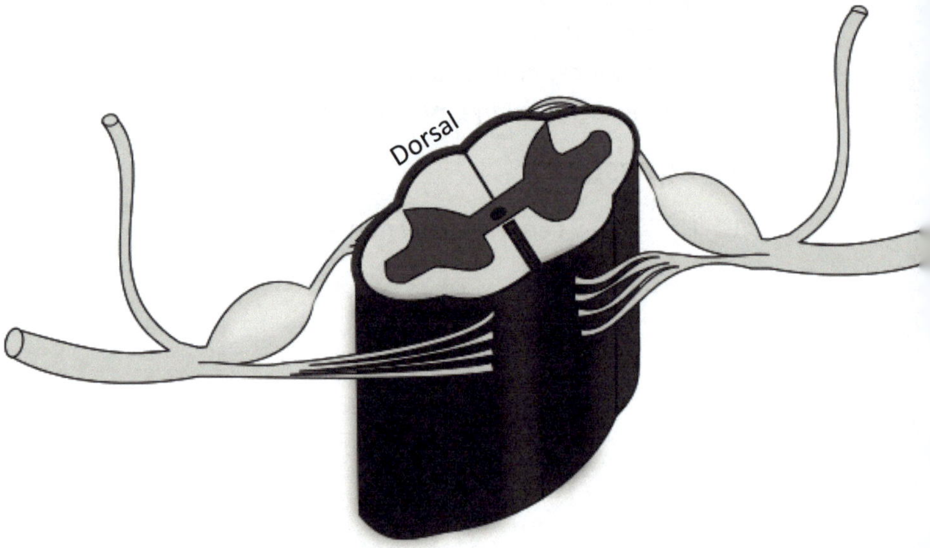


[2] "Cervical vertebra " by debivort is licensed under CC BY-SA 3.0 Unported

[3] "Spinal Cord Closeup - Cord PNG" by FlyClipArt is licensed under CC BY-SA 3.0

## Review questions:

1. What is a ganglia in general?

2. What two structures meet to form a nerve?

3. Not shown here, what 3 layers are surrounding the spinal cord?

## Application questions:

1. Redraw the H or "butterfly" shape of the spinal cord below. Label which parts are responsible for sensory input and which parts are responsible for motor outputs.

## Exercise 3: Cow spinal cord dissection

In the previous exercises you learned about the spinal cord from images and models. In this exercise you will have the chance to dissect a spinal cord from a cow. This will be an opportunity to review major structures related to the mammalian spinal cord. Remember, because of **evolution** and the field of **comparative anatomy** structures between all mammals are very similar.

To start, grab the following materials:

- Scalpel with blade
- Scissors
- Tweezers
- Gloves
- Dissecting tray with rubber mat
- One spinal cord

The following will guide you along your dissection, please follow the directions below carefully!

1. In your group, choose one person who will not wear gloves or touch the spinal cord to answer questions in this packet and read steps out loud as you go along. The spinal cord is covered in preservatives so please wear gloves if handling the spinal cord.
2. Observe the spinal cord, what is surrounding the spinal cord?

3. Take a pair of scissors and forceps and slowly start to dissect away the meninges. Use a pointed probe for the *pia mater*. In your own words, describe each of the meninges



4. The *dura mater* and *arachnoid mater* should be fairly easy to remove compared to *the pia mater*. Why?

5. Using your scalpel cut a half inch piece of the spinal cord. Using this and the rest of the spinal cord, identify structures from Exercise 2.
6. Once done, take a small piece of notebook paper and write the names of your group members on it. Place this on your dissecting tray and take a picture. On the course website, submit one of these pictures per group.

**Your dissection is complete! Do the following to get cleaned up**

1. Discard of any solid tissue in the biohazard bin
2. Rinse all tools and trays and clean with a small amount of soap
3. Dry the tools and trays with paper towels and place them back in the correct area
4. Spray your bench with disinfectant and wipe it down with a paper towel

## Exercise 4: Major spinal nerves

Originating from the ventral rami are the plexi and spinal nerves. There are numerous spinal nerves of the human body as many branch out. As you might expect, functionally, spinal nerves allow for sensation and motor control. In fact, as you learned about muscles you also learned some spinal nerves. In all, you should be to identify and explain basic information about the following plexi and spinal nerves:

- Plexi
    - Cervical
    - Brachial
    - Lumbar
    - Sacral
- Spinal Nerves
    - Axillary
    - Radial
    - Musculocutaneous
    - Ulnar
    - Median
    - Genitofemoral
    - Femoral
    - Obturator
    - Saphenous
    - Sciatic
    - Tibial
    - Common fibular
    - Pudendal

## Review questions:

1. Fill out the table below that summarizes the plexi.

| Plexi | Spinal Cord Origin (e.g. $C_1$-$T_{12}$) |
|---|---|
| Cervical | |
| Brachial | |
| Lumbar | |
| Sacral | |

2. Fill out the table below that summarizes the major spinal nerves of the human body.

| **Spinal Nerve** | **Plexus** | **Spinal Cord Origin (e.g. C1-T12)** | **One thing innervated by this nerve (e.g. muscle, specific region of skin, organ, etc.)** |
|---|---|---|---|
| Axillary | | | |
| Radial | | | |
| Musculocutaneous | | | |
| Ulnar | | | |
| Median | | | |
| Genitofemoral | | | |
| Femoral | | | |
| Obturator | | | |

10

| | | | |
|---|---|---|---|
| Saphenous | | | |
| Sciatic | | | |
| Tibial | | | |
| Common fibular | | | |
| Pudendal | | | |

10

## Exercise 4: Gross anatomy of the brain and brain stem

Below are various diagrams of the human brain. Using the list of terms below, label the following structures. Please note that not all structures will be found in each picture. To help orient you, the central sulcus is marked by an arrow for the external view.

- Frontal lobe
- Parietal lobe
- Temporal lobe
- Occipital lobe
- Cerebellum
- Pons
- Medulla oblongata
- Primary somatosensory cortex/area (post-central gyrus)
- Primary motor cortex/area (pre-central gyrus)
- Longitudinal fissure
- Oblique fissure
- Lateral sulcus
- Transverse fissure
- Arbor vitae
- Corpus callosum
- Cingulate gyrus
- Thalamus
- Hypothalamus
- Pituitary gland (not part of nervous system)
- Pineal gland (not part of nervous system)
- Lateral ventricles
- Choroid plexus
- Third ventricle
- Fourth ventricle
- Cerebral cortex

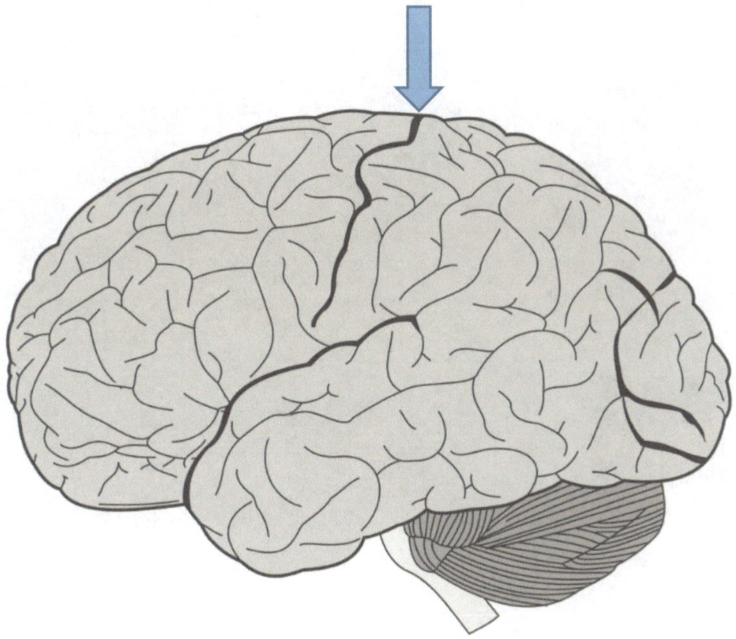

10

[4] " Diagram of human brain" by Hguiney is licensed under CC BY-SA 3.0. Recolored.

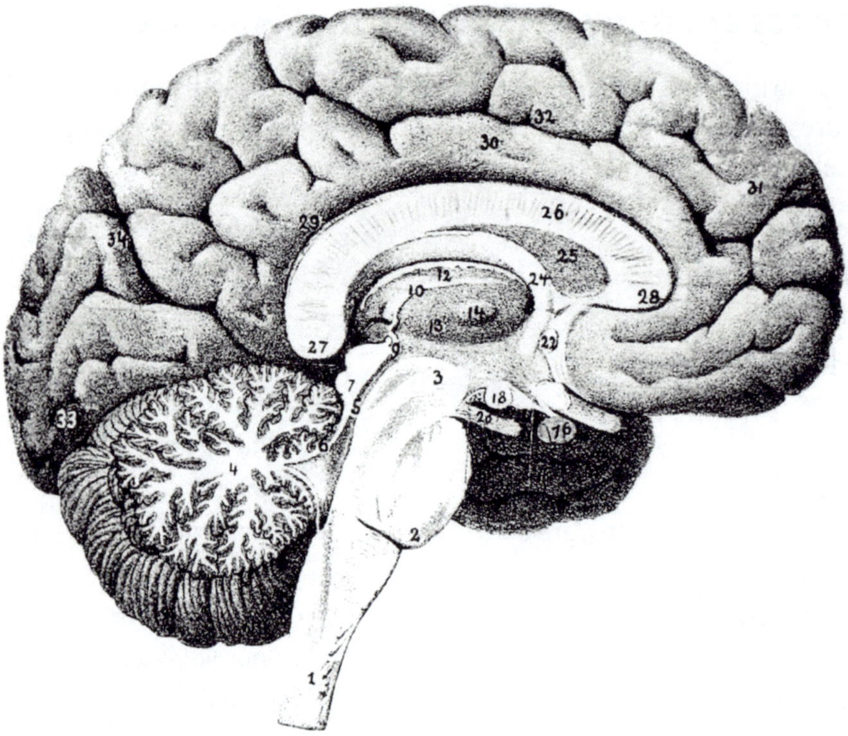

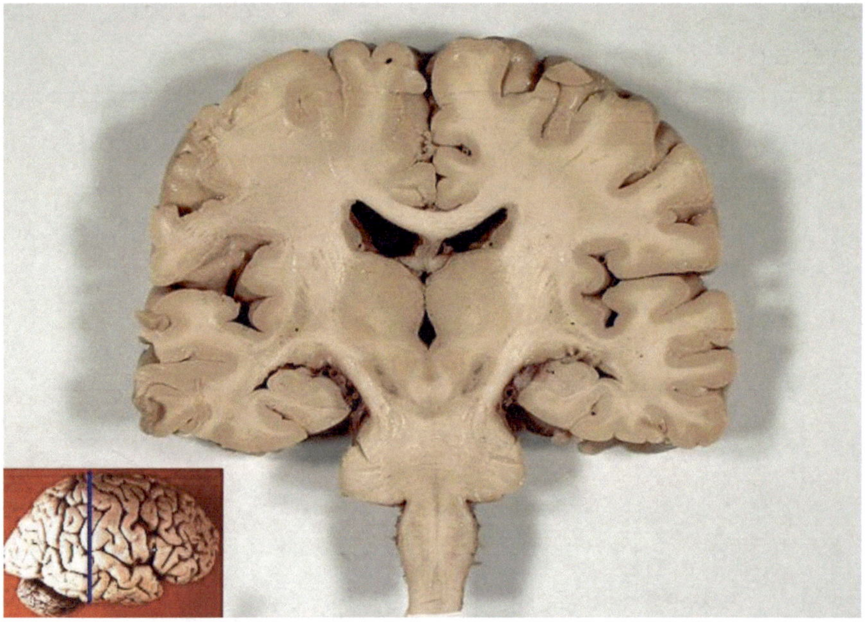



[5] Popular Science Monthly Volume 46. Public Domain.

[6] " Human brain frontal (coronal) section" by John A Beal, PhD is licensed under CC BY 2.5

## Exercise 5: Human brain dissection

In this dissection you will view a dissection of the fixed, human brain. To start, begin the video here. As you watch the dissection, answer the following questions.

1. What structure connects the two hemispheres? What type of fibers (white or grey) compose this structure?

2. The cerebellum is embryologically part of what brain stem structure?

3. What structure separates the parietal and occipital lobes?

4. What "lobe" spans the frontal, parietal, and temporal lobes?

5. The third ventricle is surrounded on either side by what structure?

## Exercise 6: Dural folds of the brain, ventricles, and meninges

The brain, like the spinal, cord is wrapped in meninges. One major difference, however, is that the dura mater is convoluted and folds inwards on the brain. This forms the dural folds, and more specifically, the falx cerebri. Explore the dural folds on any relevant models in the class.

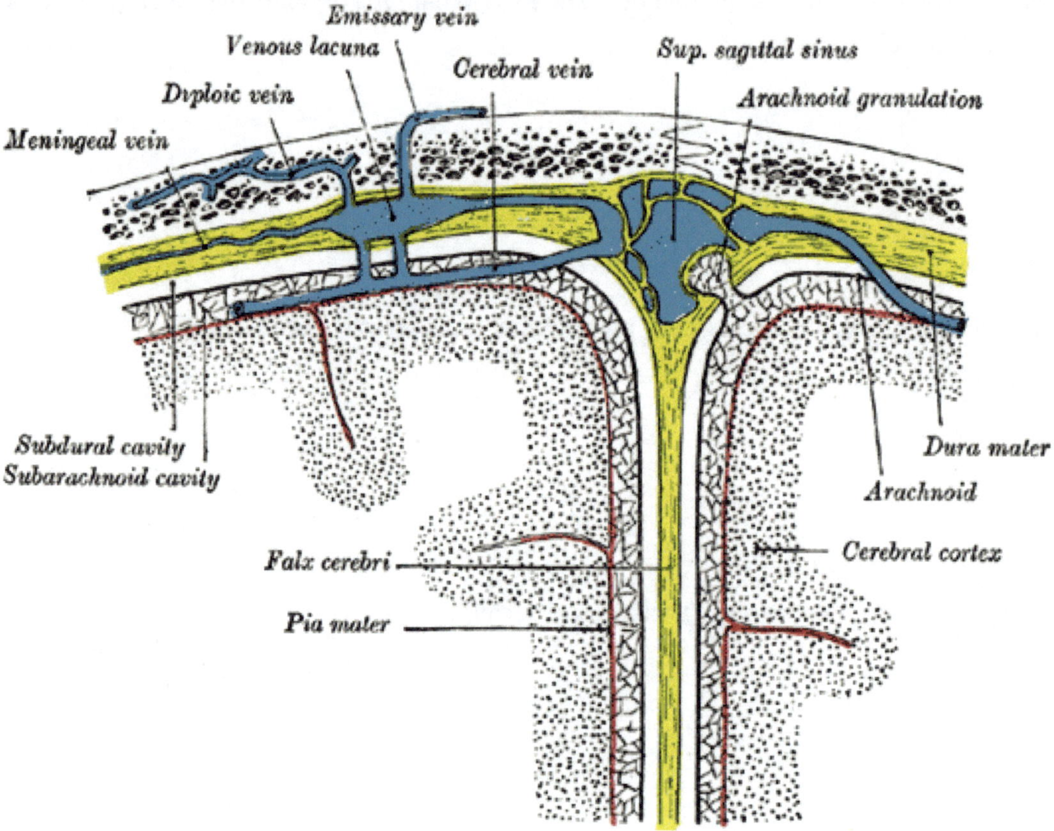


10

The dura mater is also an important structure in the brain because it houses the superior sagittal sinus. This is the point by which cerebrospinal fluid (CSF) can reenter general circulation.

### Review questions:

1. In what dural space do you find CSF?

2. What capillaries provide the brain with blood that eventually becomes CSF?

### Application questions:

1. Based on the image above, how would a bacteria reach your brain through a cut on your scalp?

2. How do you think CSF circulation would be affected if the meninges were swollen due to inflammation (as in meningitis)?

## Every-body is Different: First Hispanic Astronaut Physician

Many people dream of becoming astronauts, but few people ever do. Futhermore, very few Hispanic people have ever been an astronaut. Serena Auñón-Chancellor (1976-) was the first Hispanic physician to go on a mission. While in space her research focused on Parkinson's, a devastating neurodegenerative disorder with no cure. The goal was to crystalize a protein to understand its structure. These protein crystals are extremely finiky, so the absence of gravity is of great benefit. Her research has helped to push the scientific knowledge of this disease one-step further.

10

## Exercise 7: Cranial nerves

Unlike the spinal nerves, cranial nerves originate from the brain and brain stem. From a physics standpoint, cranial nerves are faster since action potentials travel a much smaller distance to reach the brain. These nerves are also highly protected since a significant portion resides within the skull. For the image below, label all the cranial nerves. On the right-hand side is a superior view of the skull that gives you can idea of what cranial nerves are passing through each foramen.

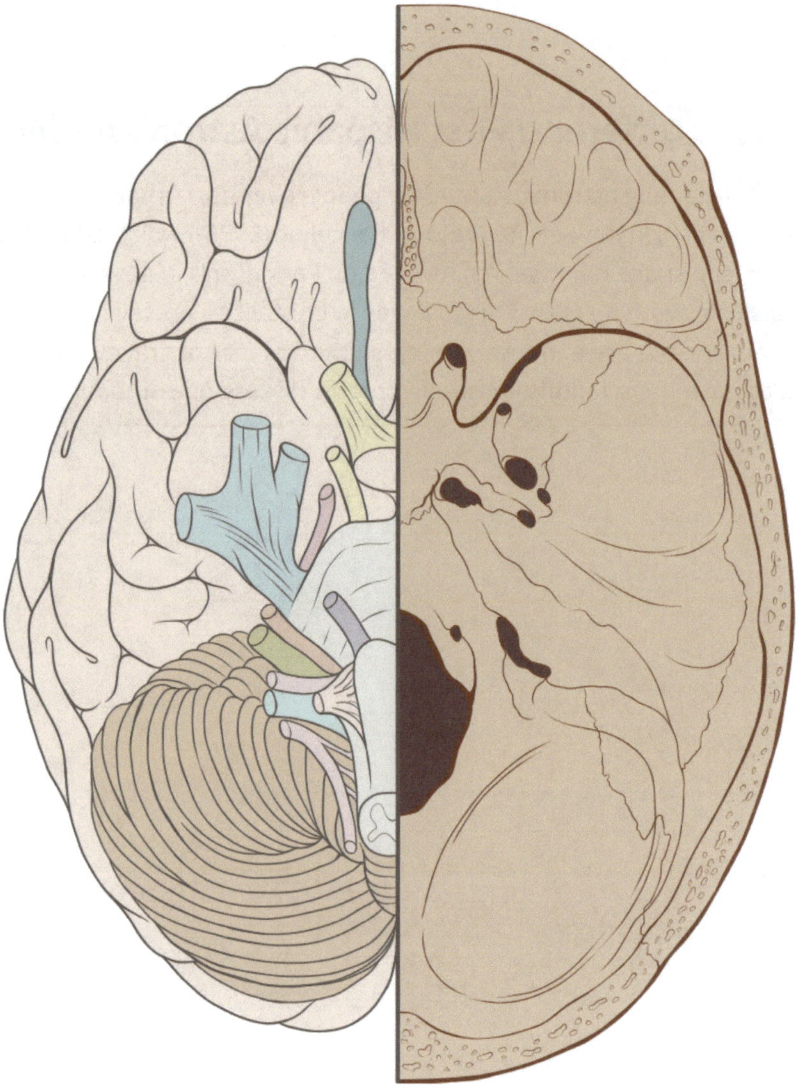

## Review questions:

1. Below is a table with various functions of cranial nerves. Name the cranial nerve that matches the function

| Cranial Nerve | Function |
|---|---|
| | Controls the sternocleidomastoid and trapezius muscles, and overlaps with functions of the vagus nerve (CN X). Symptoms of damage: inability to shrug, weak head movement. |
| | Innervates the lateral rectus, which abducts the eye. |
| | Innervates the levator palpebrae superioris, superior rectus, medial rectus, inferior rectus, and inferior oblique, which collectively perform most eye movements. Also innervates the sphincter pupillae and the muscles of the ciliary body. |
| | Innervates the superior oblique muscle, which depresses, abducts, and intorts the eyeball. |
| | Mediates sensation of sound, rotation, and gravity (essential for balance and movement). More specifically, the vestibular branch carries impulses for equilibrium and the cochlear branch carries impulses for hearing. |
| | Provides motor innervation to the muscles of the tongue (except for the palatoglossal muscle, which is innervated by the vagus nerve) and other glossal muscles. Important for swallowing (bolus formation) and speech articulation. |
| | Provides motor innervation to the muscles of facial expression, posterior belly of the digastric muscle, stylohyoid muscle, and stapedius muscle. Also receives the special sense of taste from the anterior 2/3 of the tongue and provides secretomotor innervation to the salivary glands (except parotid) and the lacrimal gland. |
| | Receives sensation from the face and innervates the muscles of mastication. |
| | Receives taste from the posterior 1/3 of the tongue, provides secretomotor innervation to the parotid gland, and provides motor innervation to the stylopharyngeus. Some sensation is also relayed to the brain from the |

| | |
|---|---|
| | palatine tonsils. This nerve is involved together with the vagus nerve in the gag reflex. |
| | Supplies branchiomotor innervation to most laryngeal and pharyngeal muscles (except the stylopharyngeus, which is innervated by the glossopharyngeal). Also provides parasympathetic fibers to nearly all thoracic and abdominal viscera down to the splenic flexure. Receives the special sense of taste from the epiglottis. A major function: controls muscles for voice and resonance and the soft palate. Symptoms of damage: dysphagia(swallowing problems), velopharyngeal insufficiency. This nerve is involved (together with nerve IX) in the pharyngeal reflex or *gag reflex*. |
| | Transmits the sense of smell from the nasal cavity. |
| | Transmits visual signals from the retina of the eye to the brain. |

# Lab 11

# Spinal Tracts

## Learning Objectives: Students will be able to...

- Identify what spinal pathway certain stimuli and actions will take
- List the steps of major spinal pathways

## Exercise 1: Spinal tracts

In short, spinal tracts are the pathway nerve fibers take to connect the brain to the spinal cord (and vice versa). Many of the nerve fibers travel through the white matter of the spinal cord, and certain regions are given names. In this class, we will cover a few of the spinal tracts. One thing to remember is that for at least motion and sensation, the pre-central and post-central gyri are the origin and ending point respectively.

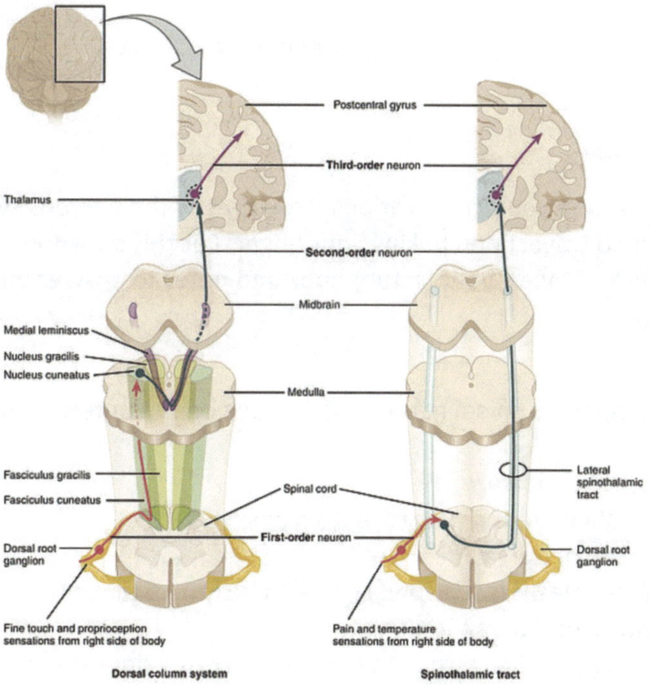


**Figure 1. Example of two different spinal tracts.**



[1] "1417 Ascending Pathways of Spinal Cord" by OpenStax College is licensed under CC BY 3.0 Unported

## Review questions:

1. Imagine you are poked on your right index finger with a needle. Place the following generic nervous system structures in the order that the nerve impulse would travel:

    1. _______Spinal nerve
    2. _______Root
    3. _______Receptor
    4. _______Rami
    5. _______Plexus
    6. _______Horn
    7. _______Cerebral cortex
    8. _______Ascending spinal tract

## Application questions:

1. In the previous question you placed various structures of the nervous system in the order a nerve impulse would travel upon poking your finger. For this question, provide specific names of the above structures. Review your lecture book and notes to answer this question See below for more information:

    a. Spinal nerve: what spinal nerve senses touch on the anterior portion of the right index finger?
    b. Root: Ventral or dorsal?
    c. Receptor: What are the pain receptors of skin?
    d. Rami: Ventral or dorsal?
    e. Plexus: What plexus is the spinal nerve (question a) part of?
    f. Horn: Ventral or dorsal?
    g. Cerebral cortex: What gyrus of the cerebral cortex does the nerve impulse end at?
    h. Ascending spinal tract: What spinal cord and brain stem structures/regions do the nerve fibers take? Note: there will be many!

2. Once you have identified the names of the specific structures for the above question place them in order based on your answer to review question 1.

## Clinical Application: Splitting the Brain

You might suspect that splitting the brain in the sagittal plane might not be good for the body, and you would be correct! However, the surgical procedure known as corpus callosotomy is actually used as a treatment for certain types of sever, untreatable seizures. In a corpus callosotomy, part or all of the corpus callosum is cut. In terms of neural pathways, this prevents the left and right hemispheres from communicating; there is no direct alternative pathway. Side effects from this including trouble speaking, cognitive impairment, and feeling as though your hands are alien. Despite these rather severe symptoms, the body is able to continue working despite the 2 hemispheres not communicating.

11

Intentionally left blank

11

Intentionally left blank

Intentionally left blank

11

Intentionally left blank

11

Intentionally left blank

11

# Lab 12

# Autonomic Nervous System

## Learning Objectives: Students will be able to…

- Give an overview of the ANS in general including function, structure, and organization
- List how the parasympathetic and sympathetic divisions differ anatomically and functionally.

## Exercise 1: Overview of the autonomic nervous system

The **autonomic nervous system (ANS)** controls aspects of our body that we are not consciously in control of. That is the, visceral aspects. This includes sweating, heart rate, digestion, etc. As a result, the ANS is a critical aspect of our nervous system. There are 2 major divisions of the ANS, the **sympathetic** and **parasympathetic** divisions. These are best described by 2 phrases, respectively: "fight or flight" and "rest and digest". The sympathetic division is activated in instances of stress or panic, and the parasympathetic in times of relaxation. For example, the sympathetic nervous system will cause your heart rate to speed up. On the other hand, activation of the parasympathetic nervous system causes your stomach to fill with acid to digest food.

Despite having 2 very different functions in the body, there are some commonalities. For example, both nervous systems have a **pre-ganglionic** neuron that synapses with the cell bodies of **post-ganglionic neurons**. A ganglia is a collection of cell bodies. Some, but not all ganglia, get special names (e.g., the celiac ganglia).The pre-ganglionic neurons for both divisions are myelinated and use the neurotransmitter **acetylcholine**. The post-ganglionic neuron for both divisions is unmyelinated.

There are also, of course, differences between these 2 division. For the post-ganglionic neuron, the neurotransmitter differs between divisions. The parasympathetic division continues to use acetylcholine, whereas the sympathetic division uses the neurotransmitter **norepinephrine** (also known as **noradrenalin**). The origins of the divisions also varies. Parasympathetic information is carried on cranial nerves III, VII, IX, and X, and spinal nerves $S_2$-$S_4$. Sympathetic information is contained on spinal nerves $T_1$-$L_2$.

Below are questions that summarize key features of the ANS.

## Review questions:

1. Fill in each blank below
   a. The sympathetic nervous system is about ___________________ and flight, whereas the parasympathetic nervous system is about rest and ___________________.
   b. Preganglionic neurons use the neurotransmitter___________________
   c. Postganglionic neurons of the sympathetic nervous system use the neurotransmitter ___________________, and postganglionic neurons of the parasympathetic nervous system use the neurotransmitter ___________________
   d. Only ___________________ neurons have myelin, ___________________ neurons don't have any.
   e. The sympathetic nervous system originates from roots of T____ to L____
   f. The cranial nerves that are part of the parasympathetic nervous system are ___________________ (4 in total)
   g. The spinal roots of the parasympathetic nervous system originate from S____ to S____

2. Draw lines to match each splanchnic nerve to its correct ganglia

| | |
|---|---|
| Lumbar splanchnic | Superior mesenteric ganglia |
| Lesser splanchnic | Inferior mesenteric ganglia |
| Greater splanchnic | Celiac ganglia |

3. Which ANS nerve innervates the adrenal glands? What is unique about this innervation?

4. The sympathetic nervous system extends into the sacral region via what structure? What structure allows the sympathetic nervous system to extend into the cervical region?

## Application questions:

1. Amphetamines are drugs that mimic neurotransmitters of the sympathetic nervous system. What neurotransmitter do you think amphetamine mimics?

2. Nicotine is a drug that mimics neurotransmitters of the parasympathetic nervous system. What neurotransmitter do you think nicotine mimics?

## Clinical Application: Heart Pills for Anxiety?

You are probably familiar with the sensation of your heart racing. Perhaps before a big presentation, or asking someone on a date, your heart might increase its beats per minute dramatically. This is due to activation of the sympathetic nervous system. While our bodies used this system to fight/flee from predators, it is still very much applicable to modern day problems. This issue is particularly tough for those who work in the performing arts. A musician, actor, dancer, etc. may only have one shot at an audition. Stage fright and nervousness could have a major impact on their livelihood.

**Beta-blockers** are a major class of drug prescribed to people with high blood pressure and other cardiovascular issues. One of its major effects is slowing down the heart. It does this by blocking the effects of the sympathetic neurotransmitter norepinephrine. However, it has been used off-label ("unofficially"; it is not approved by the FDA for this use) as an anti-anxiety drug. Its same use for blocking the sympathetic nervous system means that it can also be used prevent the anxiety-inducing effects of the sympathetic nervous system. This can be used by performing artists to reduce the jitteriness and fast heart rate that they may experience in high-stress situations such as auditioning for a new gig.

# Lab 13

# General and Special Senses



## Learning Objectives: Students will be able to…

- Describe the overall structure of the tongue
- Identify major structures of the ear (incl. middle and inner)
- Describe the surface anatomy of the human eye
- Identify major structures of the eye on a diagram, human eye model, or preserved cow eye
- Describe the overall structure of olfaction

## Exercise 1: Anatomy of the special senses

Your nervous system by itself is a collection of neurons. Its actions are not spontaneous, but instead a response to "the environment". It is through your **senses (receptors)** that your nervous system gathers the information it needs to respond. For example, if your hand is sitting on a hot surface, your body will respond by causing your biceps brachii to contract which pulls your hand away. Similarly, if you spot a bear, your nervous system will activate the sympathetic division of the autonomic nervous system. Among many things, this will cause release of epinephrine (also known as adrenalin) which directs blood flow to your muscles and speeds up your heart; both useful for getting away quickly from a predator.

The above examples highlights the 2 divisions of your senses: the **general senses**, and the **special senses**. The former has no dedicated organ or specialized region of the body that only processes that sensation. Pain (**nociception**) is a good example of a general sense. There is no single organ dedicated to pain sensation; We can sense pain in many places of our body. For special senses, there is a dedicated organ. The primary function of your eyes is sight. Similarly, the primary function of your ears is hearing and balance. In total you have 5 special senses: vision, taste, smell, hearing, and equilibrium (balance). Please note that touch is **not** a special sense. It is a general one

Since there are no dedicated organs for general senses, they are less homogenous in structure. We will instead turn our focus to the special sensory organs.

## Sense of Smell

Label the following structures on the image.

- ☐ External nares
- ☐ Nasal cavity
- ☐ Nasal concha and turbinates
- ☐ Olfactory bulb
- ☐ Olfactory foramina
- ☐ Nasopharynx
- ☐ Hard palate
- ☐ Teeth and lips
- ☐ Sphenoidal sinus
- ☐ Frontal sinus

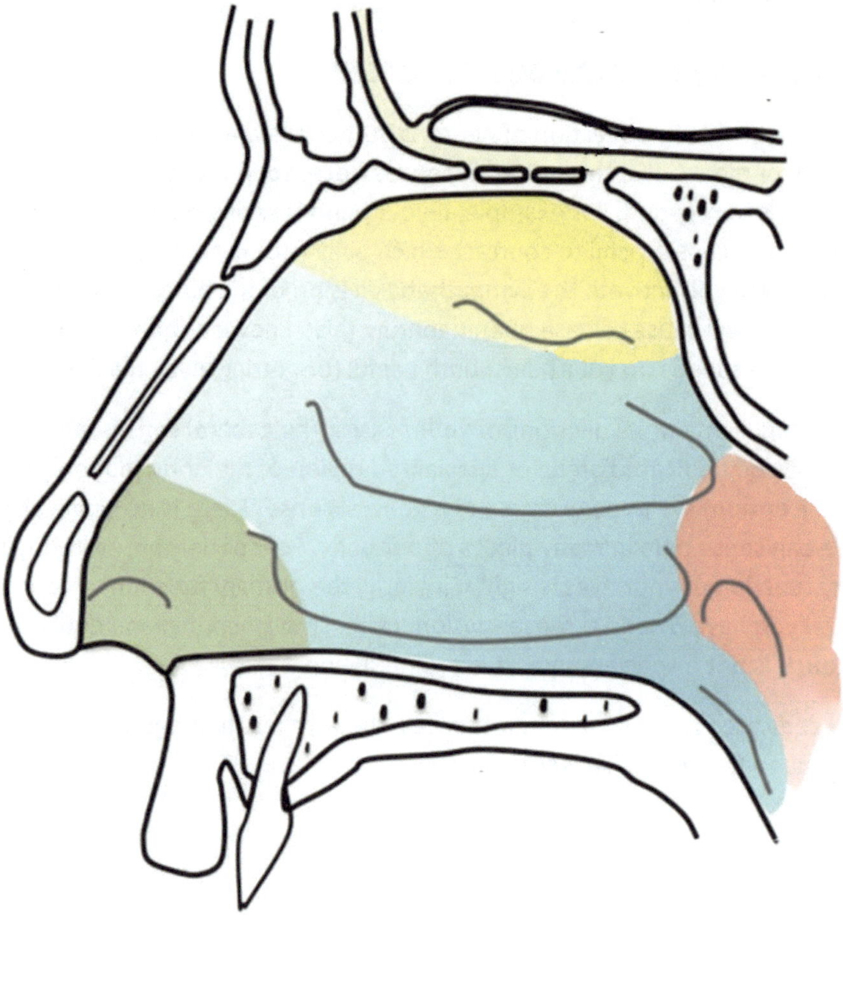

[1] "Nose diagram" by Jason Librande is licensed under CC BY-SA 4.0 / A derivative from the original work

## Sense of Vision

Label the following structures on the images below. You may have to label an item multiple times depending on the view being shown in the image.

- ☐ Eyelashes
- ☐ Conjunctiva
- ☐ Lacrimal sac (not shown)
- ☐ Cornea
- ☐ Sclera
- ☐ Iris
- ☐ Pupil
- ☐ Ciliary body with ciliary muscle
- ☐ Choroid
- ☐ Lens
- ☐ Suspensory ligaments
- ☐ Fovea centralis as part of macula
- ☐ Retina
- ☐ Optic disc
- ☐ Optic nerve
- ☐ Anterior cavity (with anterior and posterior chambers; filled with aqueous body)
- ☐ Posterior cavity (filled with vitreous humor)
- ☐ **Extrinsic (extra-ocular) eye muscles**
  - ☐ lateral rectus muscle
  - ☐ medial rectus muscle
  - ☐ inferior rectus muscle
  - ☐ superior rectus muscle
  - ☐ superior oblique muscle
  - ☐ inferior oblique muscle

### Every-body is Different: The Window to the Soul

In Western cultures, eyes are often the first thing that are noticed. Eye-contact plays a major role in non-verbal communication, and can communicate lots of useful information. It is perhaps for this reason that they are coined as "the window to the soul". Of course, eye contact also shares some other information- your eye color.

The iris is made of muscle tissue. These muscles can contract and relax to change the diameter of the pupil. However, covering this muscle tissue is a layer of stratified squamous epithelium. As you likely remember from the integumentary system, melanocytes can be found in this type of tissue. That is true for the iris as well. These melanocytes produce melanin, the concentration of which determines eye color.

For people with brown eyes (>50% of the world population) the melanocytes there are producing lots of melanin. For people with hazel, green, or blue eyes, their melanocytes are not producing as much melanin. Through scattering and refraction of light, the vast array of eye colors in humans is possible. While this can produce some striking colors, the ultimate purpose of this melanin is to protect our eyes from UV rays. Like with skin color, lighter eye colors are found in Northern European people.

Melanin production is determined by several genes, so inheritance of eye color is complex. Parents with the same eye colors can have offspring with a completely different eye color.

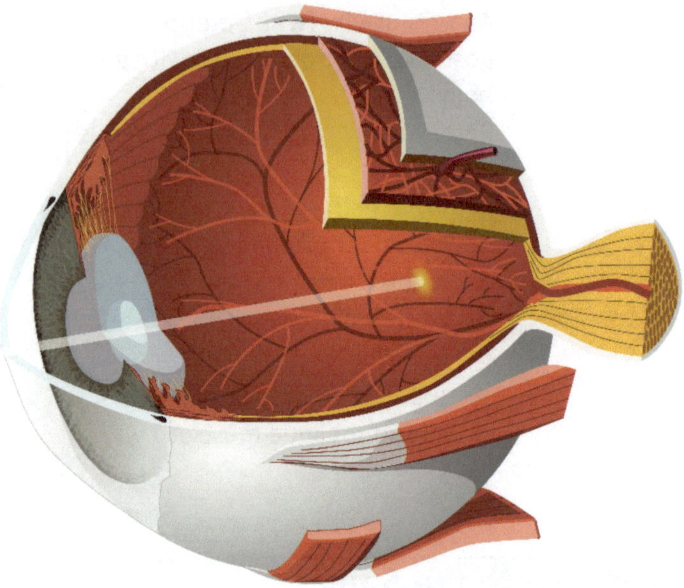

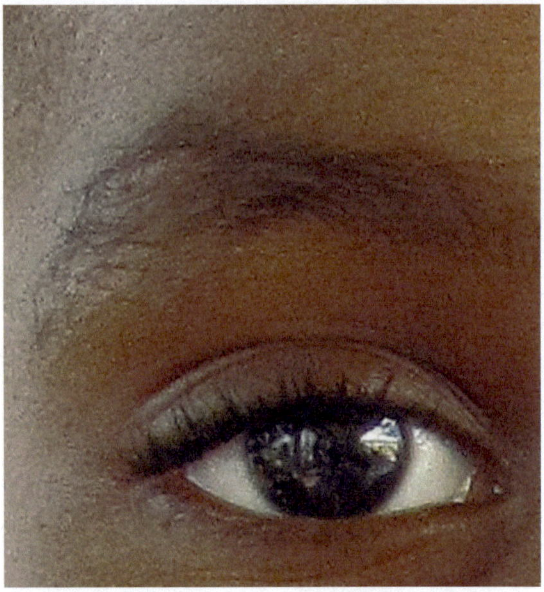

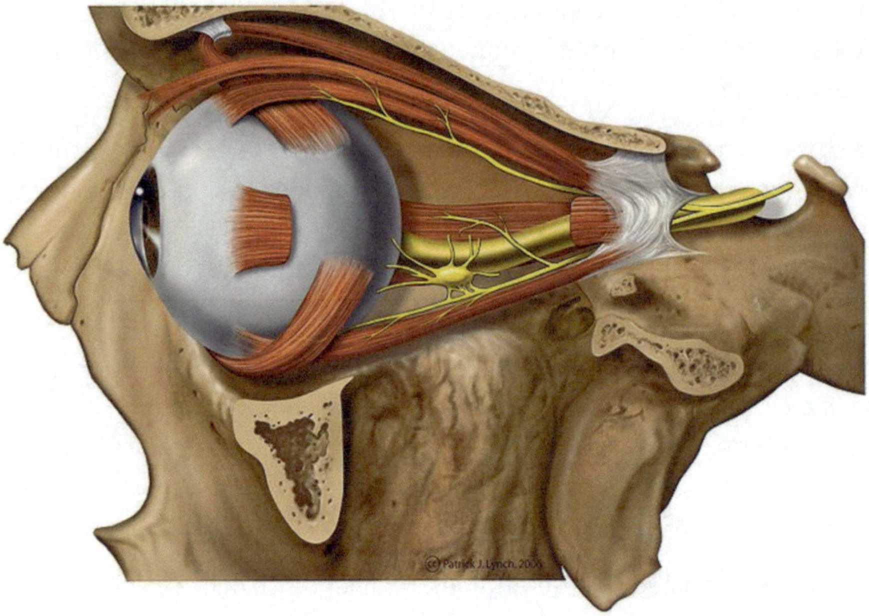


[2] "Eye-diagram.svg labels removed" by Jason Librande is licensed under CC BY-SA 4.0 / A derivative from the original work

[3] " Lateral eye and orbit anatomy with nerves" by Anka Friedrich is licensed under CC BY 2.5 Generic / A derivative from the original work

## Sense of Hearing and Equilibrium

Label the following structures on the image below. View these structures on any available models.

- ☐ Auricle/pinna
- ☐ External auditory meatus/canal
- ☐ Tympanic membrane
- ☐ Malleus
- ☐ Incus
- ☐ Stapes
- ☐ Auditory /Eustachian tube
- ☐ Vestibule (composed of saccule and utricle)
- ☐ Ampulla
- ☐ Semicircular canals/ducts
- ☐ Cochlea
- ☐ Oval window (not directly visible on models since base of stapes fills window)
- ☐ Round window
- ☐ Vestibulocochlear nerve/CN VIII
- ☐ Cranial cavity

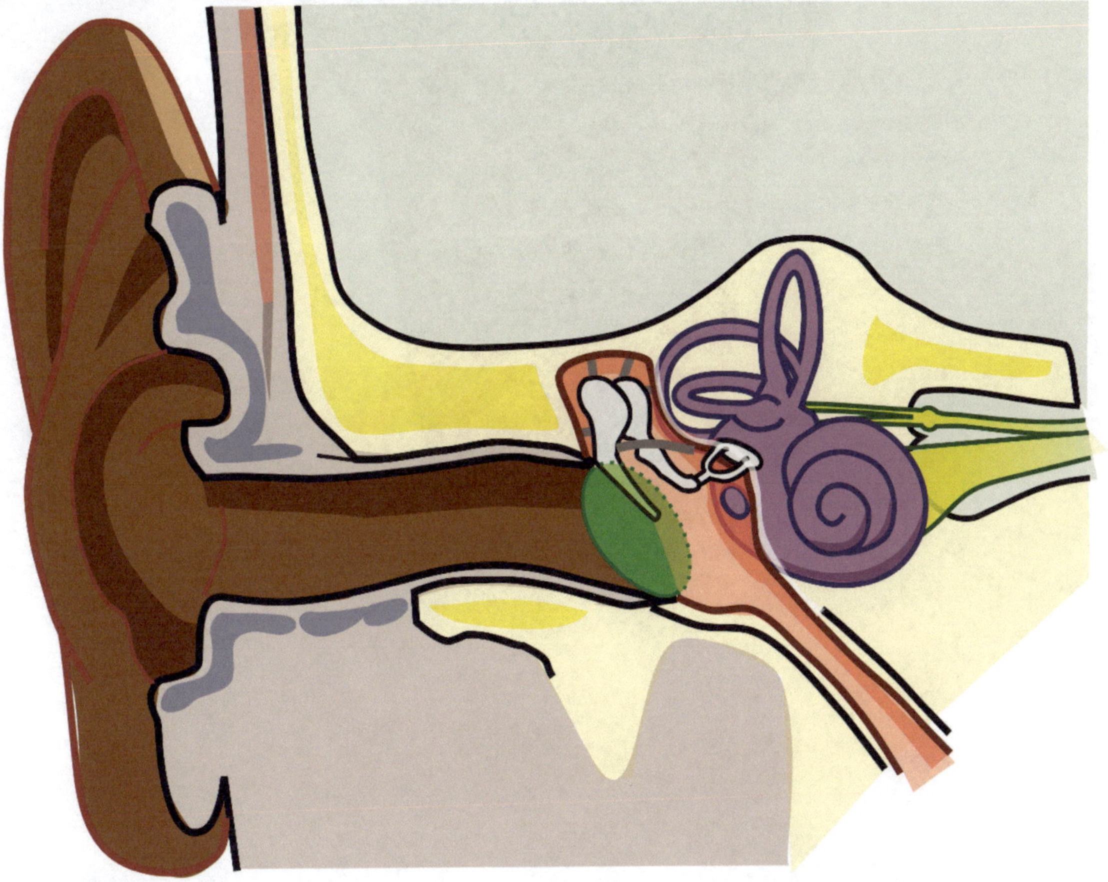

13

[4] "Ear diagram" by Jason Librande is licensed under CC BY-SA 4.0 / A derivative from the original work

## Sense of Taste

Label the following structures on the image below. View these structures on any available models.

- ☐ Circumvallate papillae
- ☐ Foliate papillae
- ☐ Fungiform papillae
- ☐ Filiform papillae (don't label, found everywhere)

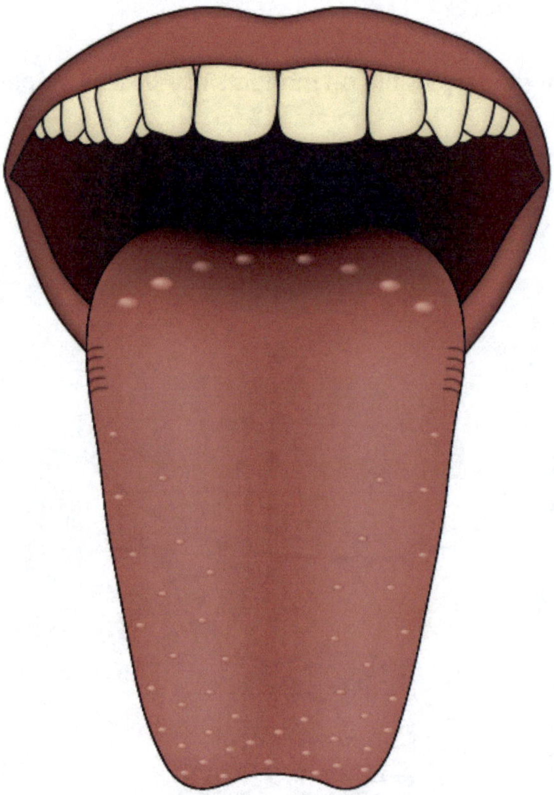

# Lab 14

# Endocrine System



**Learning Objectives: Students will be able to…**

- Identify the major, primary endocrine glands on a model or diagram, and give a brief explanation of that gland's overall function
- Explain, briefly, what each of the major hormones do

## Exercise 1: Major endocrine organs

If we were to use an analogy, the nervous system is like an airplane. It is quick, but has limited amount of cargo (information) it can transport. On the other hand, the endocrine system may be more akin to a train. It is slower, but it can transport a lot of cargo (information). It is perhaps because of this that large, systemic changes such as puberty, regulation of metabolism, and activation of the "fight or flight" response are driven by the **endocrine system** rather than the nervous system. It should also be noted that the amount of cargo is proportional to the duration of information transmission. In other words, the nervous system has information that is transmitted quickly, but also disappears quickly. The endocrine transmits information slowly, but it lasts longer.

This "information" of the endocrine system is what is called a **hormone**. These are small molecules often built off of amino acids or cholesterols that can bind to **receptors**. The receptors are either embedded in the cell membrane, or found in the cytoplasm of the cell. The receptors can interpret the information and convert it into something understandable by the cell. For example, receptors for the male sex hormone, testosterone, can instruct muscle cells to build more protein. The effect of testosterone seen at the organismal level then is an increase in muscle mass. In the broadest sense, a hormone can be thought of as a signal. Without a receptor to receive the signal, the signal will have no effect. For example, the liver does not have testosterone receptors, so testosterone does not have an effect.

The endocrine system, unlike the nervous system, is not localized to any single location in the body. Your pancreas, for example, is found in the abdominal cavity. On the other hand, your pituitary gland is part of the cranium, and your thyroid is in the anterior neck.

Below are **some** of the endocrine organs of the human body. Label each on the diagram below.

1. Hypothalamus
2. Pituitary gland
3. Pineal gland
4. Thyroid gland
5. Parathyroid glands
6. Adrenal glands
7. Pancreas
8. Ovaries
9. Testes

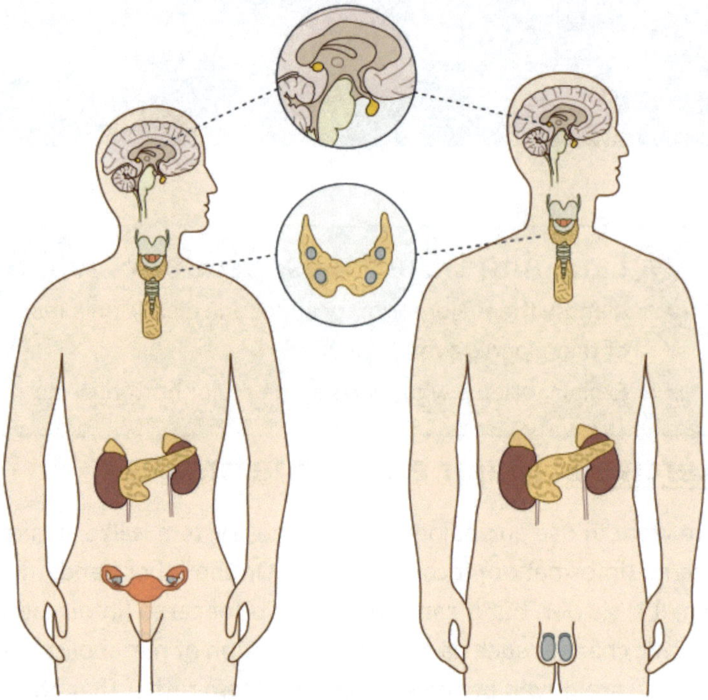

## Every-body is Different: Race and Type 2 Diabetes

Type 2 diabetes is a disease where the receptors for the hormone **insulin** are not working properly. Insulin is produced by your pancreas in response to high blood sugar, especially after you eat. It allows for channels to open in the plasma membrane that allow glucose to enter the cell. Without insulin and its receptor working properly, your cells will be unable to get glucose inside and produce the energy it needs to survive. In extreme instances this can lead to damage to the heart and blood vessels, and kidneys, blindness, and infections, among many other serious complications.

The statistic that Black people are twice as likely to develop Type 2 diabetes and Hispanic people 10% more likely, is probably something you are familiar with. The natural inclination is to think that there is a genetic difference at the heart of this. What is a more likely hypothesis is the inability to access healthy foods. Especially due to historical and ongoing systemic racism and issues of social justice, people of color have continuously been disadvantaged in terms of income. Type 2 Diabetes is often caused by unhealth diets high in fat and sugars. It is not hard to see that a meal at McDonald's cost a lot less per Calorie than a meal from Whole Foods. On top of that, there are a lot more McDonalds' in the U.S. than there are healthy grocery stores. So not only is money a barrier to a healthy diet, but so is access.

---

## Review questions:

1. Below are three columns. One contains endocrine organs, the other has hormones and the last has functions of hormones. Draw lines in between each column to connect these items. Note: there will be multiple lines extending from some of the endocrine organs since they produce multiple hormones.

| | | |
|---|---|---|
| | T3,T4 | Increases metabolism |
| Pineal gland | Parathyroid hormone | Activates fight and flight response |
| Pancreas | Cortisol | Anti-stress/inflammation hormone |
| Adrenal glands | Epinephrine | Important regulator of sleep |
| Thyroid gland | Insulin | Breaks down hydroxyapatite of bone |
| Parathyroid gland | Glucagon | Signals cells to release glucose |
| | Melatonin | Signals cells to store glucose |

Intentionally left blank

# Lab 15

# The Cardiovascular System

## Learning Objectives: Students will be able to…

- Describe the organization of the cardiovascular system
- Identify, compare, and contrast major cell types found in blood
- Identify structures of the heart
- Identify major arteries and veins of the human body

## Exercise 1: Overview of the cardiovascular system

The simplest way to describe the cardiovascular system is as a pump connected to a series of tubes. The pump is the heart, and the tubes are the blood vessels. The heart is actually two pumps. One that pumps deoxygenated blood to the lungs, and another that pumps oxygenated blood to the body. That leads to the purpose of the cardiovascular system – to transport oxygen around the body, and to collect carbon dioxide waste. This transport occurs in the blood, a type of fluid connective tissue

Blood is complex mixture of dissolved molecules composing your **plasma** (e.g., sodium, potassium, antibodies, albumin, etc.), cells (white and red blood cells), and cell parts (platelets). Perhaps the most important part of your blood is the **red blood cell** or **erythrocyte**. It is packed full of the protein **hemoglobin**. It is able to bind oxygen which allows for its transport through the blood. This is not to discredit the role of **white blood cells**, or **leukocytes**. They protect our bodies from foreign pathogens. However, they are much more abundant in the lymphatic system than in the cardiovascular system.

In terms of the vessels containing the blood, there are two major types – **arteries** and **veins**. The former brings blood away from the heart, and the latter brings blood back to the heart. Connecting these two vessels together are **capillaries**. It is the capillary where gas exchange actually happens. This ability to facilitate gas exchange is due to the capillaries being "leaky"; the lack of strong cellular junctions allows for easy transport of molecules. The arteries and veins do not have that ability. Their role is simply in the transport of blood, and not the gas exchange itself.

Finally, we have the **heart**. It is more or less, a giant muscle. However, unlike your biceps brachii where maybe you can do 10, 20, 30 etc. bicep curls, your heart works continuously throughout your life. In fact, it will beat about 2.5 billion times in your life. That is definitely more than you could ever hope to do with skeletal muscle tissue! Just like skeletal muscle, your heart is coordinated by the nervous system. It has the ability to speed up or slow down its beating. One major difference from skeletal muscle though, is that the heart has **autorhythmicity**. Even without the nervous system, the heart can continue to beat. This is due, in part, to the gap junctions that exist between cardiomyocytes. This allows for action potentials to jump from cardiomyocyte to cardiomyocyte which facilitates beating of the heart.

## Every-body is Different: First Open-Heart Surgery

Heart surgery even today is an incredibly risky procedure. Imagine being one of the first to perform it! Daniel Hale Williams (1856-1931) was a Black physician working in Chicago during the late 1800s, only one of a few at the time. He is considered to have performed the first successful open-heart surgery when he repaired the pericardium of a patient. Outside of this impressive achievement, he also established the first interracial teaching hospital in Chicago.

## Review questions:

1. Fill in each blank below
    a. The pump of the cardiovascular system is the ____________________, the tubes connected to the pump are ____________________ and ____________________
    b. ____________________ bring blood away from the heart, and ____________________ bring blood to the heart
    c. The fluid pumped by the heart is called ____________________. It is a type of ____________________ tissue
    d. The 4 blood types are: ____________________.
    e. ____________________ artery send blood to the lungs, and the ____________________ vein brings blood back
    f. Heart muscle is made up of ____________________ muscle tissue
    g. ____________________ acts as gap junctions in between cardiomyocytes
    h. The strongest chamber of the heart is the ____________________

15

# Exercise 2: Cardiovascular histology

Blood carries a variety of cells and nutrients as it travels throughout our bodies. In this section we will review major cells found in the blood including erythrocytes and various white blood cells. Some time will also be spent on the vessels carrying blood: the arteries and veins.

## Arteries and Veins (25)

View this slide on the microscope. It will contain both arteries and veins for you to view. The easiest way to differentiate between the two will be based on thickness. Arteries are thicker since they need to deal with high blood pressure. Beyond that, the layers that compose each are the same.

**Label the following**: Artery and vein (may need multiple drawings), tunica intima (may not be visible), tunica media, tunica adventitia, lumen



Total Magnification:

## Blood Smear (26)

This slide will be an opportunity to view blood under the microscope. You will see a variety of cells, but most importantly you should focus on locating the five white blood commonly found in blood. These will likely not be found close together, so you have been provided multiple circles below for your drawings. In fact, you will see that these white blood cells are dwarfed by the number of red blood cells.

Total Magnification:



Total Magnification:

Total Magnification:

Total Magnification:

Total Magnification:

15

## Review questions:

1. The most abundant cell type in blood is/are...

2. Based on a hematocrit, what percentage cells on the blood smear that you viewed are white blood cells?

3. The below table compares the major white blood cells covered in class. For this table, provide the name of each white blood cell on your own.

15

| Name of WBC | Relative size (large/small) | Shape of nucleus | Granules (yes/no) | Color |
|---|---|---|---|---|
| | | | | |
| | | | | |
| | | | | |
| | | | | |
| | | | | |

# Exercise 3: The heart

Label the diagrams to below with the following structures. While doing this, view any models, cadaveric specimens, or other resources available to you.

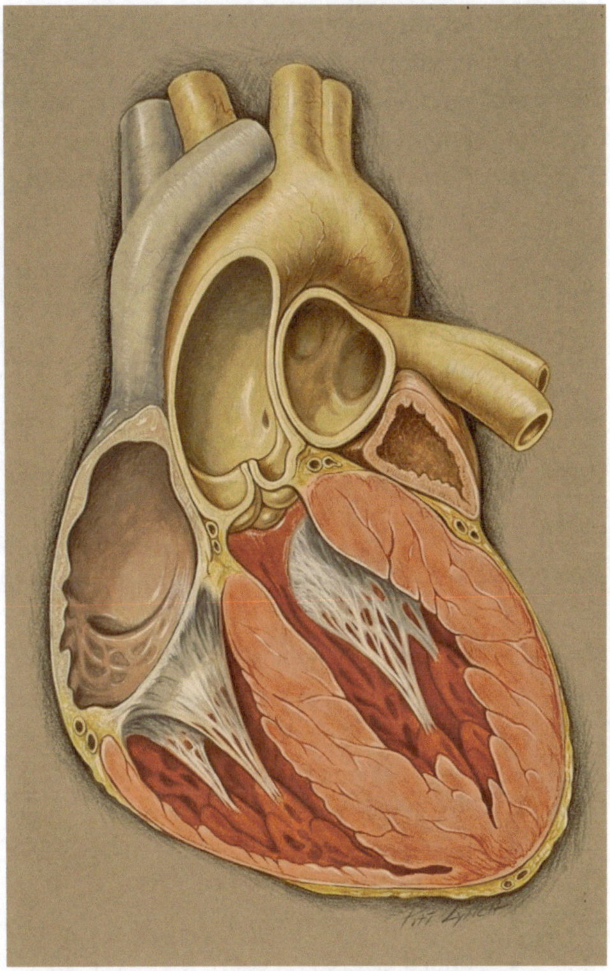

- Left atrium
- Right atrium
- Left ventricle
- Right ventricle
- Valves (aortic, right AV [tricuspid], pulmonary and mitral not shown)
- Pulmonary artery
- Pulmonary vein (not shown)
- Aorta
- Trabeculae carneae
- Chordae tendineae
- Papillary muscle
- Superior vena cava
- Inferior vena cava (only hole is shown in RA)
- Septum
- Apex



Label the diagram of the heart to the right:

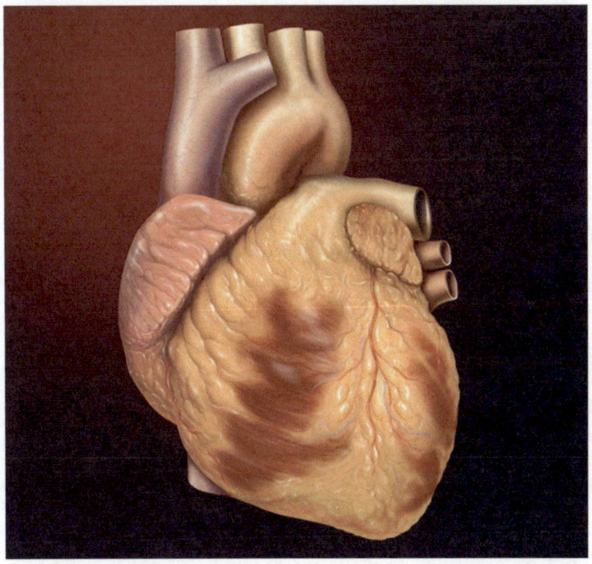

- Left anterior descending artery (LADA)
- Auricles of atria
- Apex
- Base
- Margins (inferior, superior, etc.)

## Exercise 4: Major vessels of the human body

Using models, cadavers, and other resources to identify the following arteries and veins.

- Superior vena cava
- Inferior vena cava
- Pulmonary artery
- Pulmonary vein
- Aorta (descending, abdominal, etc.)
- Subclavian artery and vein
- External carotid artery
- Internal carotid artery
- External jugular vein
- Internal jugular vein
- Celiac trunk
        - Left gastric artery
        - Common hepatic artery
        - Splenic artery
    - Superior mesenteric artery and vein
    - Inferior mesenteric artery and vein
- Inferior phrenic artery and vein
- Suprarenal artery and vein
- Renal artery and vein
- Gonadal artery and vein
- Hepatic portal vein
- Hepatic vein
- Axillary artery and vein
- Brachial artery and vein
- Ulnar artery and vein
- Radial artery and vein
- External iliac artery and vein
- Internal iliac artery and vein
- Femoral artery
- Great saphenous vein
- Vertebral artery and vein
- "Circle of Willis"
    - Anterior cerebral artery
    - Anterior communicating artery
    - Internal carotid artery
    - Posterior cerebral artery
    - Posterior communicating artery

# Lab 16

# The Lymphatic System

## Learning Objectives: Students will be able to...

- Give a general outline of lymphatic anatomy including major lymph nodes, structure of a lymph node, etc.
- Explain what "lymph" is
- Compare and contrast blood, interstitial fluid and lymph

## Exercise 1: Overview of the lymphatic system

Capillaries are inherently leaky. Ions, gases, and even small proteins can leak from them. When compounded with blood pressure, this leakiness is a generally one-way street. Molecules can only leak *out* of capillaries, not back in. In turn, the body needs a way to return the fluid leaked by the capillaries (now called **interstitial fluid**) to your cardiovascular system. They do this by entering into lymphatic vessels. The fluid is now called **lymph**, but it is very similar to the composition of blood plasma minus larger molecules.

Most of the lymphatic vessels of the body do not have names, but one exception this is the **cisterna chyli**. It is a large abdominal lymphatic vessel that conducts lymph from the lower half of your body upward. It should be noted that the lymphatic vessels have no pump. They rely on gravity, contraction of skeletal muscle, and very rarely smooth muscle to propel the lymph. This is one reason why putting your feet up after a long day feels so good! It allows for lymph to more easily circulate. When lymph does not circulate, this leads to **edema**. For example, if you stay on your feet for too long (e.g., working in a retail job), you might develop mild edema. In fact, an easy experiment to do is to measure the circumference of your ankles before and after work. Eventually, the lymph makes it back to low pressure veins of the body such as the subclavian vein. Through this pattern of circulation, the cardiovascular system can maintain a fairly constant blood volume.



Beyond that, the lymphatic system houses many parts of your **immune system**. While the cardiovascular system does house many white blood cells, the vast majority of your white blood cells. The **lymph nodes** (the major organ of this system) is often compared to "filters" in that they trap pathogens. The white blood cells then attack these pathogens and release numerous inflammatory molecules. Palpating your lymph node is one easy way in which doctors can determine if you have a serious infection.

Using circles and arrows, label the approximate location of the following lymphatic system structures.

- Submental lymph nodes
- Axillary lymph nodes
- Inguinal lymph nodes
- Spleen
- Cisterna chyli
- Thymus
- Tonsils

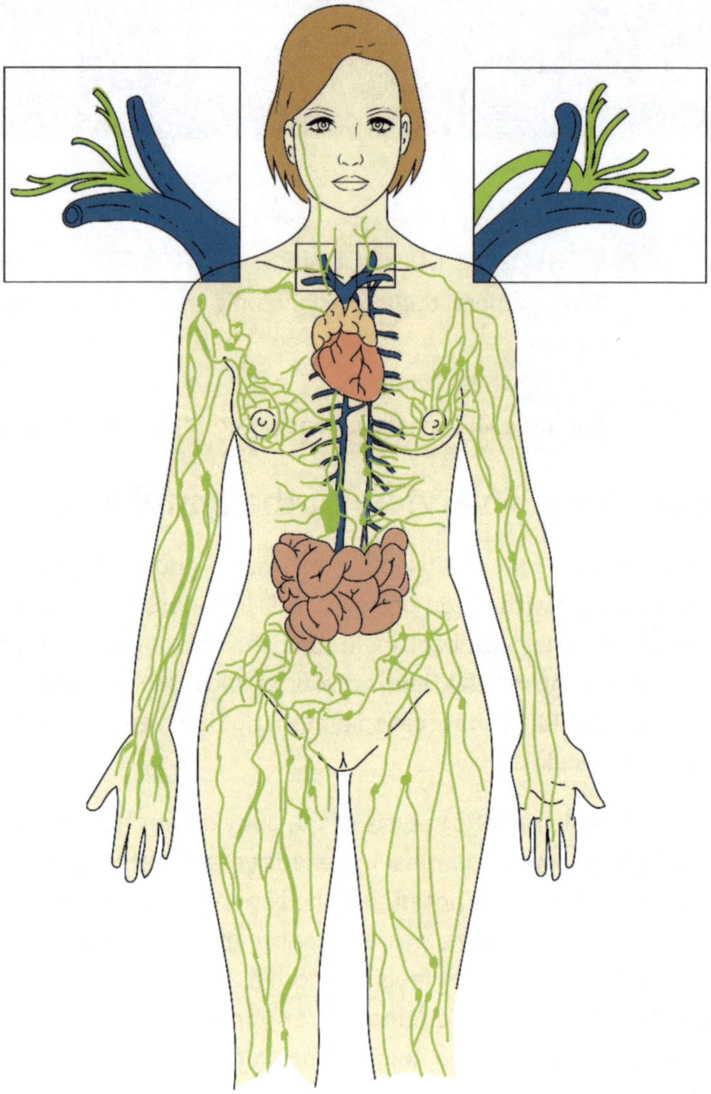

## Review questions:

1. Fill in each blank below to summarize the lymphatic system:
    a. When blood leaves the capillaries, it becomes ________________. When that fluid then enters the lymphatic vessel, it is called ________________
    b. The majority of lymphatic vessels in the lower half of the body converge at a lymphatic sac called the ________________
    c. Your ________________ removes erythrocytes and acts similarly to a lymph node
    d. The ________________ is most often palpated by doctors to determine if you have an infection.

2. What are the three or four tonsils of the human body? What are the functions of tonsils?

3. Why is it hard to find your tonsils if you are an adult?

4. Through what veins does lymph return to the cardiovascular system? What veins are responsible for which regions of the body?

## Application questions:

16

1. Elephantiasis is a disease caused by a parasite that infects the lymphatic system. It produces symptoms as shown on the right. Propose a mechanism for how the symptoms in the picture arise

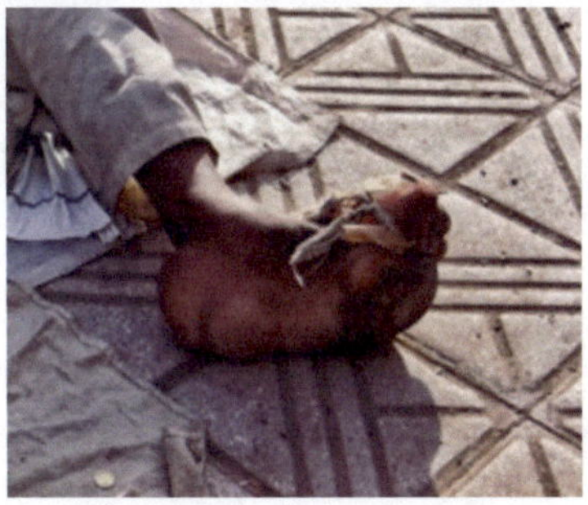

2. Comatose patients need to be rotated occasionally by nurses. Explain from the perspective of the lymphatic system why this is important.

[1] " Elephantiasis in Herat " by Peretz Partensky 

## Exercise 2: Lymph nodes

The lymph node is a major organ of the lymphatic system, and has a critical role in its immune function. Inside of it there are numerous T-cells, B-cells and other immune cells that are continuously monitoring the lymph as it passes through. Label the following structures on the lymph node below.

- Afferent lymphatic vessel
- Valves (of vessels)
- Efferent lymphatic vessel
- Hilus
- Capsule
- Cortex
- Follicle
- Germinal center

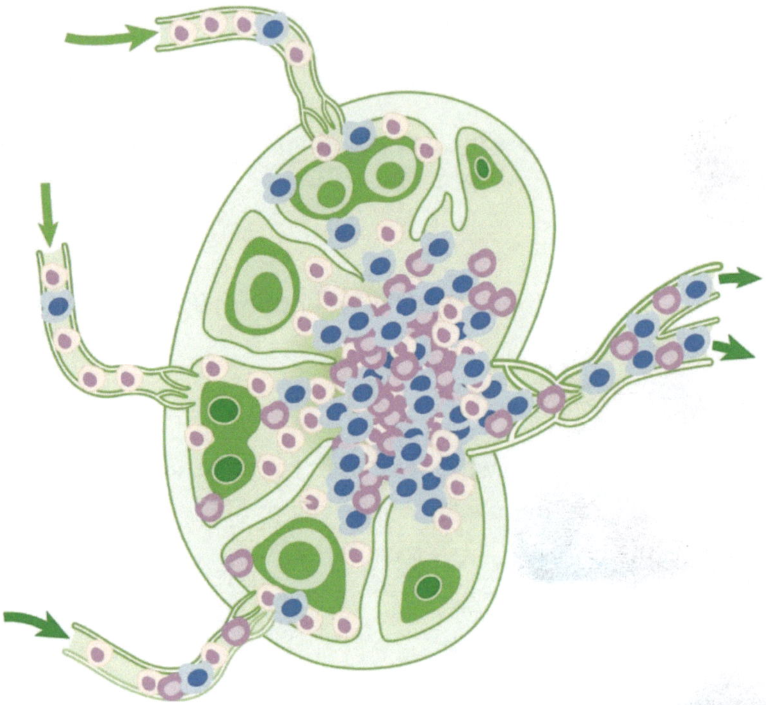

16

[i] " Lymphatic Trunks and Ducts " by ASCCC OERI is licensed under CC BY 4.0

[2] "Unlabeled diagram showing the structure and function of lymph nodes." by Andrewmeyerson is licensed under CC BY-SA 4.0

# Lab 17

# The Respiratory System

## Learning Objectives: Students will be able to...

- List and identify the structures of the Respiratory system from the nose to alveoli
- List the 3 types of cells found in alveoli
- List and identify major structures used in vocalization

## Exercise 1: Respiratory system gross anatomy

The **respiratory system** is closely tied to the cardiovascular system. Without the respiratory system red blood cells would not be able to pick up oxygen and deliver carbon dioxide. These molecules are critical for ensuring that our cells can create the energy they need to survive.

The principal organ of the respiratory system is the **lungs**. In total, you have 2 lungs with a total of 5 lobes. That may seem like an odd number considering our bodies are fairly symmetrical, but the heart poses a problem. It shares space in the thoracic cavity, and occupies quite a bit of the left half of the body (the heart is a left pointing organ). Because of this, the left lung has a space called the **cardiac notch**. The heart fits snugly into the left lung which has only 2 lobes, not 3 like the right lung.

Comprising your lungs are million of tiny sacs called **alveoli**. It is here that gas exchange actually occurs. The alveoli are lined with simple squamous epithelial tissue which flat shape allows for efficient diffusion across the cell membrane. Unlike the capillaries of your cardiovascular system, which are leaky, the alveoli form a continuous barrier. This highlights the intimate balance between the need for thinness, and protection from pathogens. Your respiratory system is exposed to the outside environment, so pathogens can travel along with the air. It is perhaps not surprising that so many microbial diseases affect our respiratory system.

The respiratory system is composed not only of your lungs, but also the tubes that connect to it (i.e. the trachea, bronchi, etc.). Helping support this tubing is a variety of connective and muscular tissue. But, before we even get to the trachea we must review the upper respiratory tract. The **nasal cavity** is the major structure of the upper respiratory tract. It warms, cleans, and moisturizes air on its way to our lungs. If you have ever been outside on a very cold day, you may know that it can hurt to breathe the air! From the nasal cavity it passes through the back of your throat (**pharynx**) then to the voicebox (**larynx**) and then into your lower respiratory tract.

Of course, we cannot ignore the respiratory system's role in vocalization. Imagine a world with no speaking or singing. It is thanks to our unique larynx, tongue, and nasal cavity that we can do what we can with our voices.

**Label the diagrams below:**

- Nares
- Mouth
- Palate
- Nasal concha and meatus
- Nasopharynx
- Opening to auditory/eustachian tube
- Frontal sinus
- Sphenoidal sinus

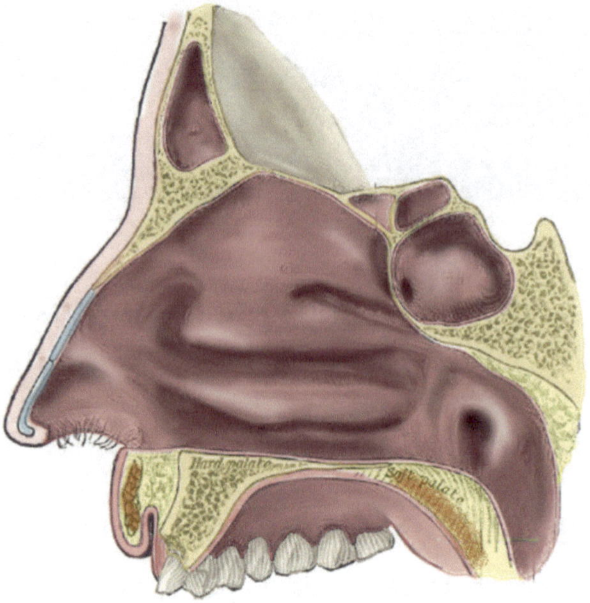


- Thyroid cartilage
- Cricoid cartilage
- Tracheal cartilage
- Carina
- Primary bronchi
- Secondary bronchi
- Tertiary bronchi
- Superior lobe of right lung
- Middle lobe of right lung
- Inferior lobe of right lung
- Horizontal fissure
- Oblique fissure of right lung
- Superior lobe of left lung
- Inferior lobe of left lung
- Oblique fissure of left lung
- Cardiac notch

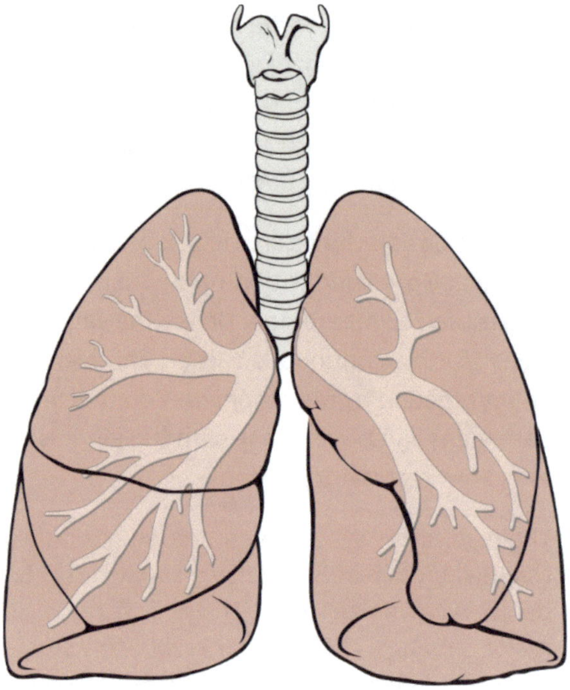



[1] "Sagittal view of the lateral wall of the nasal cavity" by Cypressvine is licensed under CC BY-SA 4.0

[2] " Lungs-simple diagram of lungs and trachea" by Patrick J. Lynch is licensed under CC BY 2.5

The below image is for reference:

**Figure 1. Alveoli of the lung**

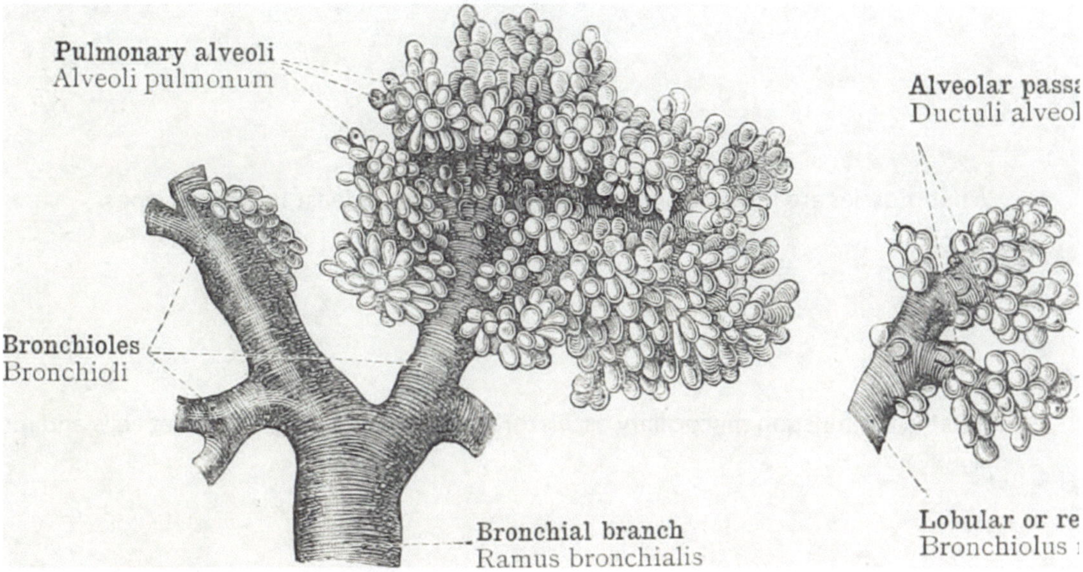


## Clinical Application: People of Color and Skin Disease

COVID-19 is known mostly for its respiratory symptoms. However, people occasionally develop issues with the cardiovascular system. This is probably best described as a wholesale increase in the risk of cardiovascular diseases. The cause for this is not entirely understood, but it is thought to have something to do with a protein called ACE2. This is a receptor found in respiratory tissue, but is also found in the cardiovascular system. ACE2 functions as a receptor and it is found embedded in the plasma membrane. COVID-19 is aware of this though, and has evolved to co-opt ACE2 to enter the cell. By binding ACE2 it can be endocytosed by your own cells and begin replicating. While this happens in the respiratory tissue in particular, it is thought that COVID-19 can enter the blood stream and replicate there. For example, it might bind to ACE2 receptors of the heart and replicate in cardiac muscle tissue. The vascularity of the lungs allows for simple diffusion, but it is also a liability for the body.

17

## Review questions:

1. How many lobes are there for each lung?

2. What muscles are responsible for inhalation? Name at least two major ones.

3. What is the function mucociliary escalator? What are the roles of goblet cells and mucous glands?

## Application questions

1. Occasionally instead of steroidal inhaler drugs which reduce inflammation during an asthma attack, a person is prescribed a sympathomimetic drug instead. What tissue is a drug mimicking the sympathetic nervous system likely affecting? How does this relate to the evolutionary purpose of the "fight or flight" response?



2. How does the above relate to an epi-pen (epinephrine injection) during anaphylactic shock?

## Exercise 2: Respiratory system histology

### Trachea (27)

View this slide. It will present you with a cross section of the trachea and esophagus. In order to orient yourself quickly, find the ciliated, pseudostratified columnar epithelium. Once you have found this you now know where the lumen ("airway") is and can work your way backwards to find other structures. Intercalated within that layer of epithelial tissue you will also find goblet cells which produce mucous. Directly beneath that epithelial tissue layer is then the *tunica mucosa*, a layer of loose connective tissue that may contain epithelial mucous glands. Underneath the tunica mucosa is a tough layer of dense irregular connective tissue called *tela submucosa*. Finally, the trachea is lined with hyaline cartilage formally known as the thyroid, cricoid or tracheal cartilages.

**Label the following:** Ciliated pseudostratified columnar epithelium, goblet cells, tunica mucosa, tela submucosa, mucous gland (if present), hyaline cartilage

Total Magnification:

## Alveoli (28)

**Label the following:** Alveolus, blood vessel

Total Magnification:

## Review questions

1. What are the three types of alveolar cells, what are their functions?

2. Type 1 alveolar cell is what type of epithelial tissue?

17

Page intetionally left blank

17

# Lab 18

# The Digestive System

## Learning Objectives: Students will be able to...

- List and identify major structures and organs of the digestive system and their functions
- List the major functions of the digestive system
- List and identify major layers found in digestive system microanatomy
- Compare and contrast the microanatomy of various parts of the digestive system
- List and identify important structure of liver microanatomy

## Exercise 1: The digestive system

At first glance, the **digestive system** and respiratory system may seem like very different organ systems. However, they share some important traits. For example, they are concerned with getting molecules across the plasma membrane. Oxygen for the respiratory system, and various nutrients for the digestive system. They are also both exposed to the external environment. Respiratory infections are common, but so are digestive ones. The major difference is that the digestive system must break down nutrients into smaller molecules that can cross the cell membrane. Oxygen and carbon dioxide being so tiny don't have that issue.

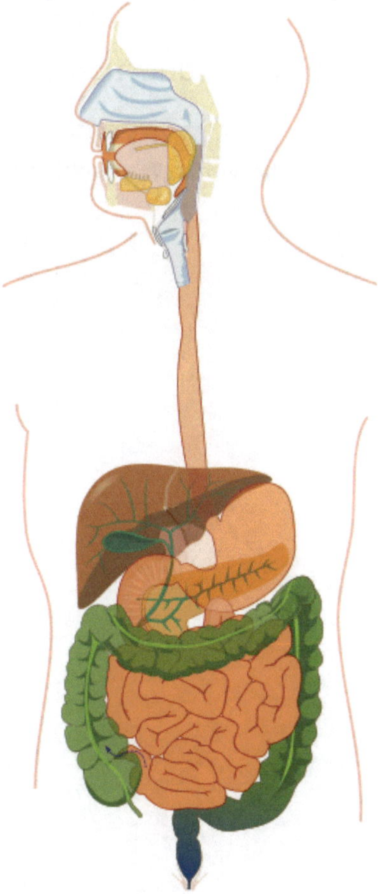

The **digestive tract** can be roughly split into five parts: The mouth, esophagus, stomach, small intestine and large intestine. However, the digestive *system* is composed of the digestive tract plus accessory organs such as the liver and pancreas and associated structures (e.g., the appendix). As a whole, the digestive system aims to break down food into small enough molecules to be utilized by the body. However, there are other functions that should not be ignored such as the digestive system's role in housing beneficial bacteria and as protection from harmful pathogens. To start, label the major portions of the digestive system below. In later exercises of this lab we will investigate each organ in greater detail.

- Mouth
- Esophagus
- Stomach
- Small and Large intesintes
- Liver
- Gall bladder
- Pancreas

## Review questions:

1. Match each organ or structure with its function

| | |
|---|---|
| Mouth | Stores bile |
| Esophagus | Produces bile |
| Stomach | Contains acid for chemically breaking down food |
| Small intestine | Absorbs majority of nutrients |
| Large intestine | Absorbs water and contains vitamin producing bacteria |
| Liver | Produces an alkaline fluid containing many enzymes to break down food chemically |
| Gall Bladder | Mechanically breaks down food with mastication ("chewing") |
| Pancreas | Transports food through the thoracic cavity |

## Application questions:

1. The digestive system can be described as a "tube open to the external environment". Explain how this statement is true. But also explain how the tube can be "closed".

## Exercise 2: The mouth

The mouth's major role is **mastication** ("chewing"). This aids in **mechanical digestion** of food by breaking it down into smaller particles. However, it is important not to forget that **saliva** also has enzymes such as salivary amylase which start the process of chemical digestion. Saliva is produced by exocrine glands called "salivary glands". In total you have six of these, three on each side of your mouth.

Label the following picture:

- Tongue
- Lips ( incl. orbicularis oris m.)
- Salivary glands (not shown, label their approximate locations)
    - Parotid
    - Sublingual
    - Submandibular
- Uvula and soft palate
- Hard palate formed by...
    - Palatine bone
    - Maxillae
- Mandible
- Vestibule (of mouth)
- Oropharynx
- Nasal cavity
- Epiglottis
- Palatine tonsils (not shown, label their approximate locations)

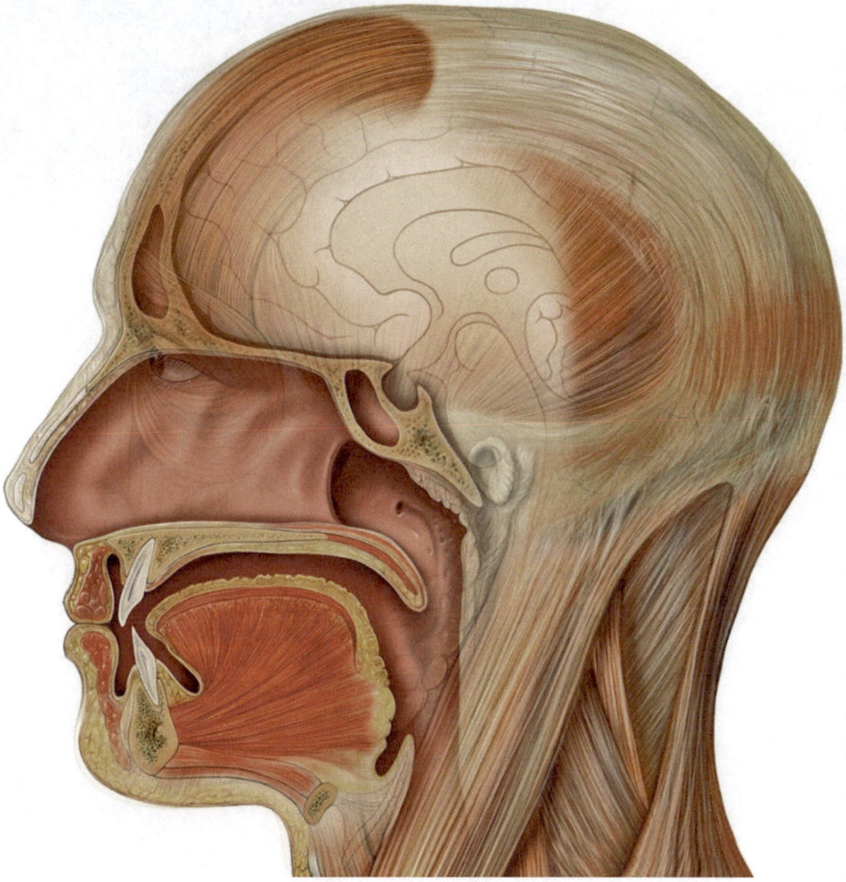

[1] "head lateral view with sagittal mouth anatomy" by Patrick J. Lynch. CC BY 2.5 Generic.

- Incisors
- Canines
- Pre-molars
- Molars

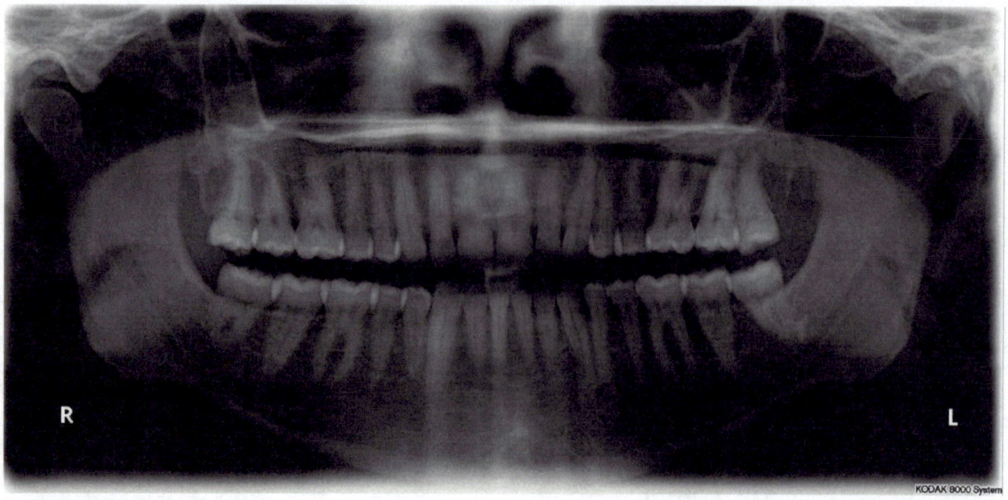


- Enamel
- Dentin
- Nerves, arteries and veins
- Apical foramen
- Mandible or maxillae
- Alveoli
- Periodontal ligament

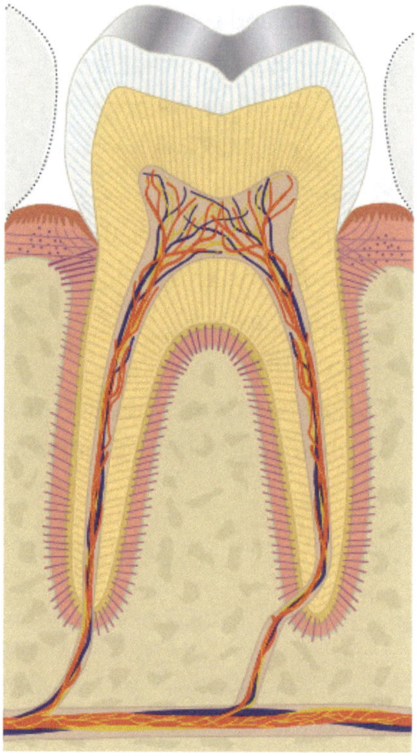

---

[2] "Panoramic X-ray of all 32 teeth of a male in his 40s with no cavities or fillings or other dental work." by Ruhrfisch CC BY-SA 4.0 International

[3] "Cross section of a tooth" by Ian Furst. Labels deleted. CC BY-SA 3.0 Unported

## Review questions:

1. What is the function of the uvula?

2. How many of each of the following do you have in one quadrant of your mouth and total

| Tooth | # in one quadrant | # total |
|---|---|---|
| Incisor | | |
| Canine | | |
| Pre-molar | | |
| Molar | | |

3. What tooth is referred to as your "wisdom tooth"?

4. What substance makes up enamel?

## Exercise 3: Esophagus and stomach

The **esophagus** acts as a tube to transport the **bolus** of food behind the heart and lungs to the stomach. From there the stomach uses both mechanical and **chemical digestion** to break down food. Because of that, the stomach is lined externally with a variety of smooth muscle, and internally with a variety of chemical producing cells.

- Esophagus
- Lower esophageal sphincter (not shown)
- Regions of stomach
    - Cardia
    - Fundus
    - Body
    - Pylorus
- Greater curvature
- Lesser curvature
- Pyloric sphincter
- Rugae (of stomach)
- Small intestine

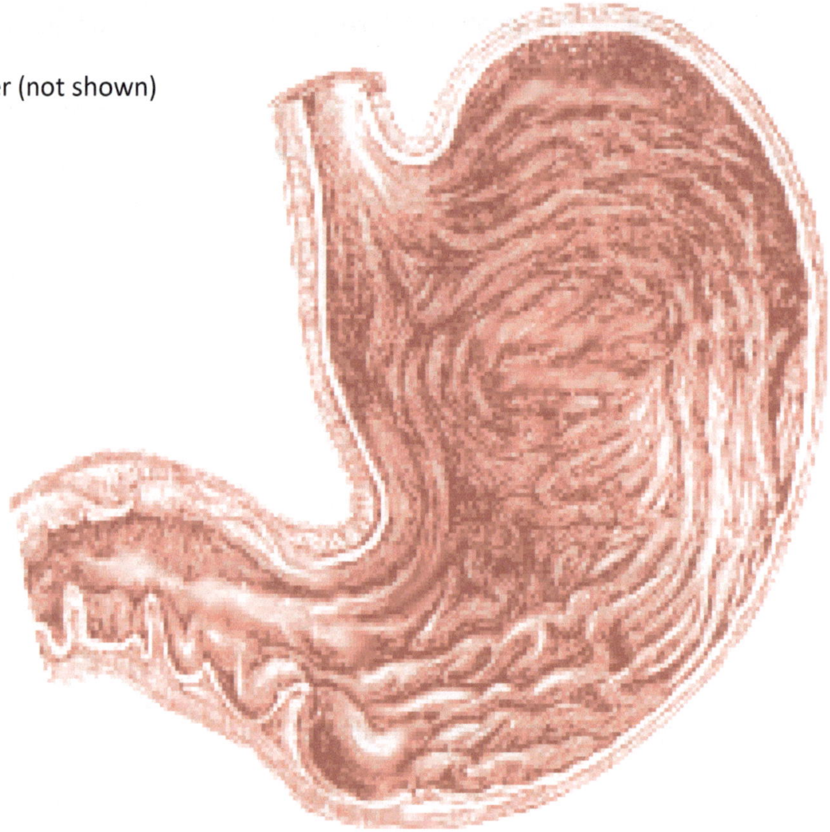

## Review questions:

1. What major structure attaches to the greater curvature of the stomach. What is the similar structure that attached to the lesser curvature of the stomach?

2. What are the three orientations of smooth muscles surrounding the stomach?

3. What disease is affiliated with weakening of the lower esophageal sphincter?

## Application questions:

1. Gastric bypass (Roux-en-Y) surgery can be used to help obese people lose weight. The diagram below shows the anatomy after the surgery. Explain how the changes below result in weight loss.

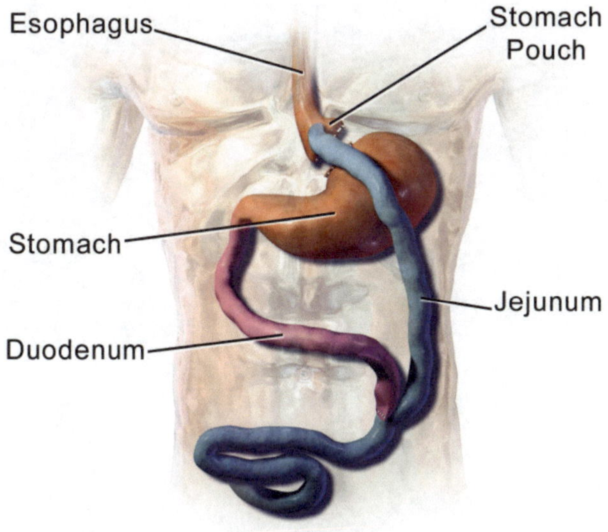


Roux-En-Y

[4] "An illustration depicting the Roux-En-Y gastric bypass procedure" by BruceBlaus. CC BY-SA 4.0 International.

## Exercise 4: Small and large intestine, pancreas and liver

After the mouth and stomach have thoroughly digested the food, the remainder of what needs to be done is absorption. This is in addition to neutralizing the stomach and preparing the nutrients for absorption, a feat accomplished by the liquids produced by the liver and pancreas.

Label the following diagram:

- Duodenum
- Remainder of small intestine (jejunum and ileum)
- Ileocecal valve
- Haustra
- Taenia coli
- Cecum
- Appendix
- Ascending colon
- Transverse colon
- Descending colon
- Sigmoid colon
- Rectum
- Anus
- Liver
- Gall bladder

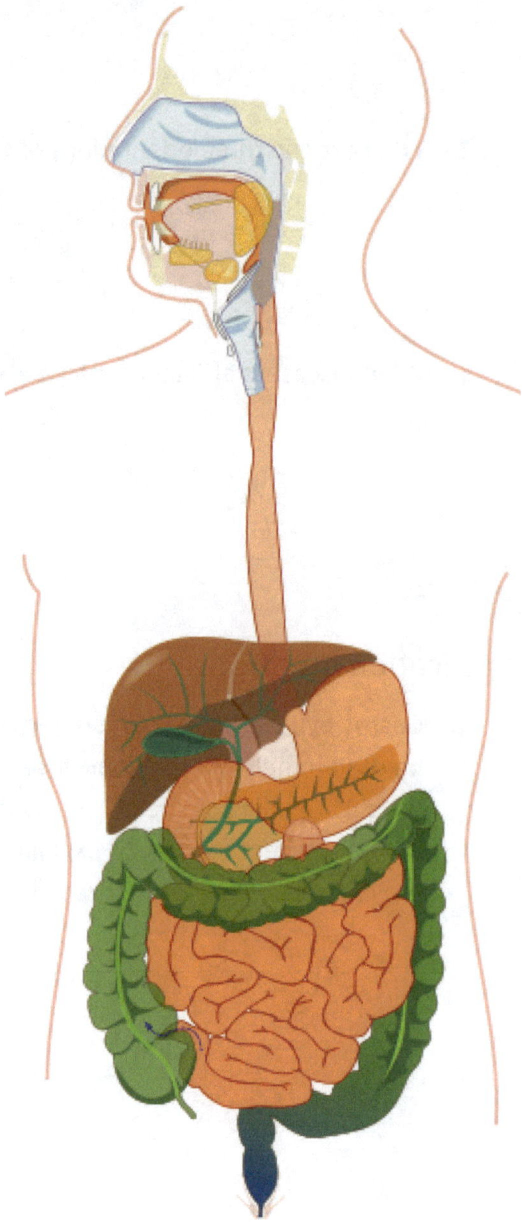

18

[5] Pixabay licsense. Labels deleted.

The below shows a posterior view of the liver and gall bladder, and their related ducts along with those of the pancreas.

Label the following:

- Left hepatic duct
- Right hepatic duct
- Common hepatic duct
- Common bile duct
- Cystic duct
- Gall bladder
- Pancreatic duct
- Falciform ligament

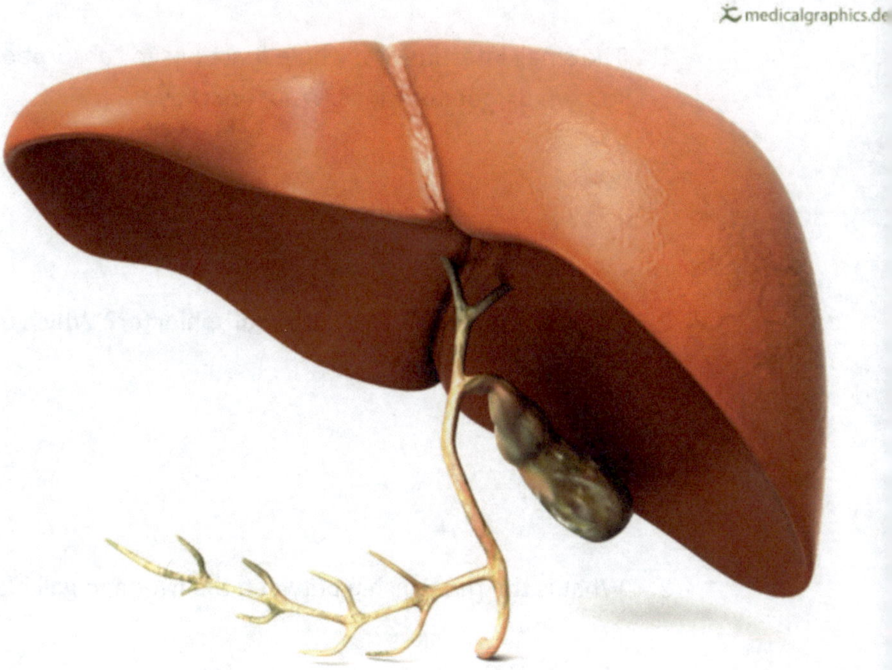


[6] "liver (back)" by www.medicalgraphics.de. 

## Review questions:

1. The small intestine as a whole functions in "absorption" of nutrients. However, each section has its own unique functions. What are they?

2. What are two muscles of the anal sphincter? Which one is made of skeletal muscle and in turn under voluntary control?

3. What is the relationship between the liver and gall bladder in terms of bile?

4. What type of gland is the pancreas (hint: it is two different types since it excretes and secretes substances)?

5. What is the difference between villi and microvilli?

6. What is a lacteal and what is its function?

7. What is the function of the hepatic portal system?

## Application questions:

1. Boxers will often try to punch with the left arm and hit the lateral, right side of their opponent. What organ are they likely targeting with their punch? Why might hitting this organ be exceedingly dangerous/painful/debilitating? Is there anything anatomically protecting this organ?

## Exercise 5: Digestive system histology

A significant number of events are happening at the microscopic level of the digestive system. Consequently, it is important to be aware of major, microscopic structures of the digestive system.

### Esophagus (29)

The esophagus follows the same basic layout as majority of the digestive system: a mucosa, followed by submuscosa, followed by a layer of external muscle and then a layer of connective tissue called adventitia (if retroperitoneal) or serosa (if intraperitoneal). Using the slide from your slidebox, draw it and then label the following structures

> **Label the following:** Lumen, mucosa, submucosa, *muscularis mucosae*, *tunica muscularis externa* (longitudinal and circular muscles), tunica serosa

Total Magnification:

## Liver (30)

The liver acts a detoxifier for the body. To accomplish this, it has numerous blood vessels providing materials to be inspected by the hepatocytes. This accomplished by a variety of enzymes that perform chemical reactions to break down chemicals.

**Label the following:** Venules, hepatocytes, sinusoids

Total Magnification:

This page intentionally left blank

# Lab 19

# The Urinary System



## Learning Objectives: Students will be able to...

- Identify major structures of the urinary system when given a diagram, model, or sheep kidney; Both at the gross anatomical, and microanatomical levels
- Trace the flow of urine through the kidney
- Match each part of a nephron with its function

## Exercise 1: The urinary system

Your body consumes a large amount of carbon, oxygen, hydrogen and nitrogen. These four elements make up about 96% of the human body's composition. For cabon, oxygen and hydrogen, these elements are used in large quantities during cellular respiration where we ultimately breathe out carbon and oxygen as carbon dioxide, and evaporate hydrogen in the form of water vapor. However, nitrogen does not follow a similar pathway. Instead, we urinate any nitrogenous waste. This waste is usually generated as proteins are broken down. For example, breakdown of muscle tissue would result in urination of large quantities of nitrogenous waste.

The urinary system's most important function is in the removal of nitrogenous waste. But, in doing so it must also conserve water. Normally this water would flow in equal quantities to the amount of nitrogenous waste. The kidneys, however, are able to concentrate this "urine" by reducing the amount of water. In turn, we are urinating gallons of water each day, but instead a relatively small amount. This is especially good for cardiovascular system. Believe it or not, but urine is derived from blood; not the digestive system. You can imagine that your blood, not intestines, are a reservoir for your body fluids. The kidneys pull water from this reservoir to form urine. In turn, there is a strong tie between blood volume and the urinary system.

19

**Label the following**

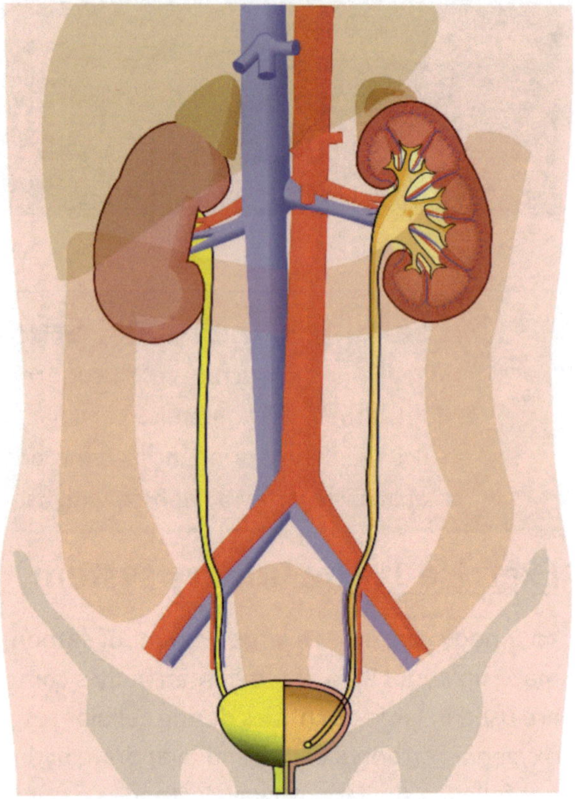

- Kidneys
- Renal artery and vein
- Ureters
- Bladder
- Urethra
- Adrenal glands

## Review questions:

1. What type of waste does the urinary system remove?
2. What is the source of the waste from question 1?
3. What is the major muscle of the bladder? What type of muscle is it?

## Application questions:

1. Assume you are urinating more than normal. What would happen to the volume of blood in your body? What would happen to blood pressure? Why does urination relate to blood?

[1] "Urinary System" by Jordi March i Nogué. Labels Removed. 

## Exercise 2: The kidneys

The kidneys by far are the most important organ of the urinary system. They have the important job of concentrating urine and regulating what substances are excreted. We will first take a look at the gross anatomy before moving on to what is happening at the cellular and molecular levels.

**Label the following:**

- Capsule
- Renal pyramid (circle and label)
- Minor calyx
- Major calyx
- Ureter
- Renal column
- Renal cortex
- Renal medulla
- Renal sinus
- Renal hilum
- Renal artery and vein

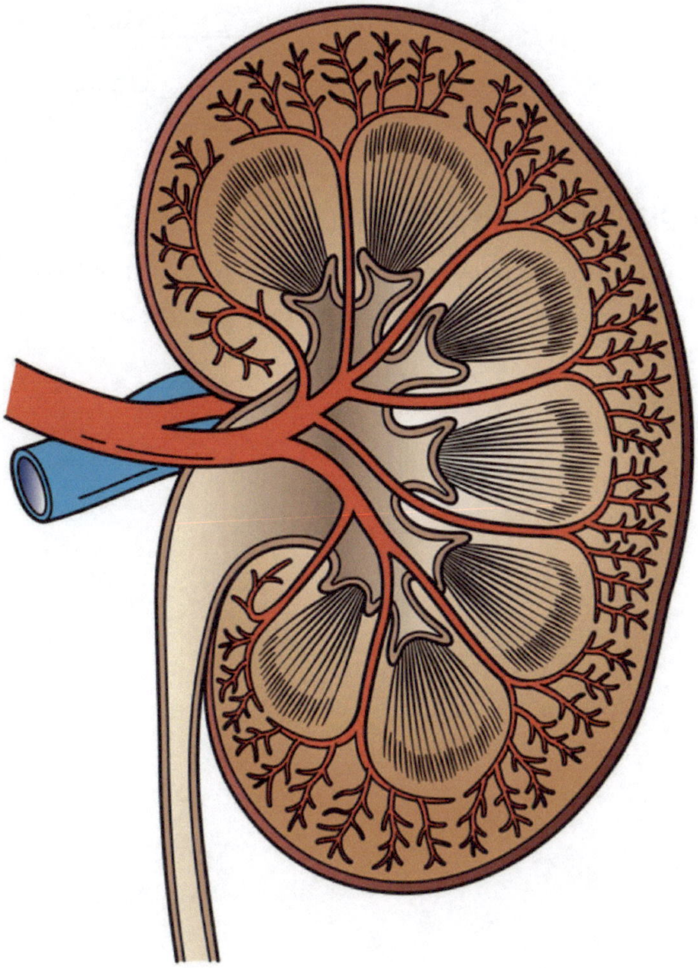

[2] "cross section of a kidney" by Holly Fischer CC BY 3.0 Unported.

## Review questions:

1. What is the correct order of the following?

_____Major calyx

_____Minor calyx

_____Ureter

## Application questions:

1. Often kidneys are removed due to cancer or to donate to someone else who needs one. What do you think the effect of removing one of your kidneys would be on blood pressure?

## Exercise 3: The Nephron

The nephron is the functional unit of the urinary system. In turn, the kidney is composed essentially of numerous nephrons. The nephron itself is a series of tubes that have cells specialized for transfer of specific molecules. For example, the descending loop of Henle is specialized for reabsorption of water.



**Label the following:**

- Glomerulus
- Bowman's capsule
- Descending loop of Henle
- Ascending loop of Henle
- Proximal convoluted tubule
- Distal convoluted tubule
- Collecting duct

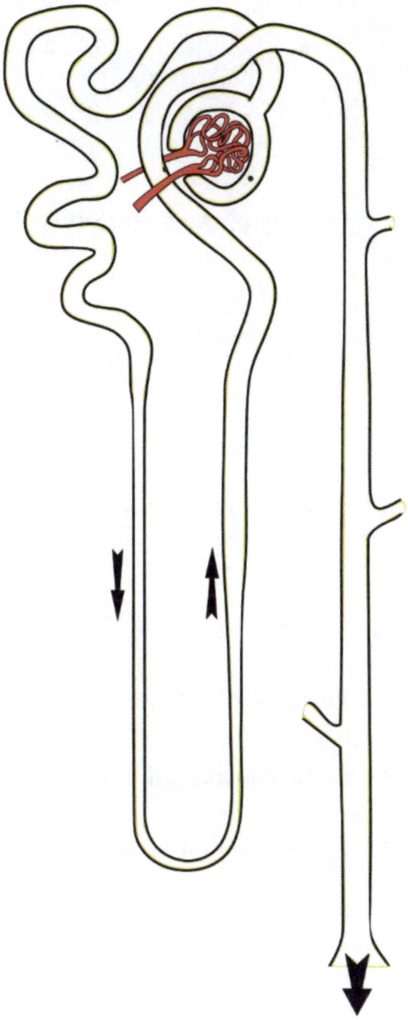

[3] "Basic anatomy of nephron" by Yosi I. Labels and shading removed. Creative Commons Attribution 3.0 Unported.

## Review questions:

19

1. What is the correct order of the following?

_____Glomerulus

_____Descending loop of Henle

_____Proximal convoluted tubule

_____Distal convoluted tubule

_____Ascending loop of Henle

_____Minor calyx

2. What is the role of the juxtaglomerular complex (not show in the image above)?

3. What molecules are transported by of each of the following sections of a nephron?

    1. Descending loop of Henle:___________________________

    2. Proximal convoluted tubule:___________________________

    3. Distal convoluted tubule:___________________________

    4. Ascending loop of Henle:___________________________

# Lab 20

# The Reproductive System and Development



## Learning Objectives: Students will be able to...

- Identify major structures of the male and female reproductive systems in cross-section
- List the 3 germ layers and what they become
- List basic facts about human reproduction
- List major steps of human development and when they happen

## Exercise 1: The three germ layers

Like the vast majority of animals, humans develop from three germ layers: ectoderm, mesoderm, and endoderm. These germ layers are thought to arise around three weeks post fertilization. This is around the stage that cells of the blastula invaginate in a process called gastrulation.

Despite all these processes happening early in development, it is important to remember that the entire human body arises from one cell (and consequently from just these three tissues)

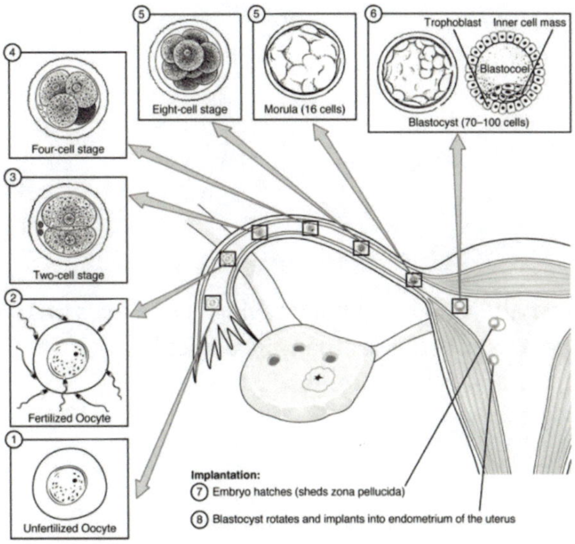


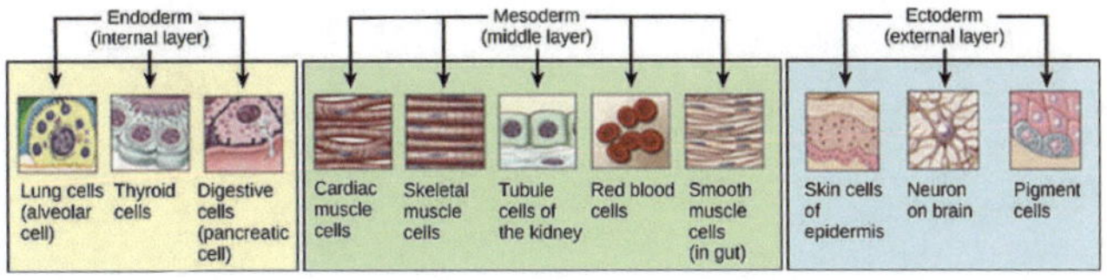


[1] "Germ layers" by CNX. CC BY-SA 3.0 Unported

[2] "Pre-embryonic development" by OpenStax.CC BY 3.0 Unported.

## Review questions:

1. Give **multiple** examples of cells and tissues that each germ layer becomes in the human body
    a. Ectoderm:
    b. Mesoderm:
    c. Endoderm:

20

2. During what week of development do these germ layers arise? Is this before or after implantation?

## Application questions:

1. The nervous system has the longest development period in the human body. Why do you think it is also the organ system most suceptible to developmental defects?

2. An infamous ad against abortion has a baby saying "I had eyes at 14 days". Based on what you know about germ layers, why is this factually incorrect?

## Exercise 2: Fetal circulation

An important aspect of fetal development is circulation. Because the baby does not have access to oxygen it must rely on the mother's cardiovascular system to provide oxygen and to remove waste. It is also not uncommon for amniotic fluid to be filled with waste as well, such as dead skin cells and even urine. Use the image below to answer the following questions.



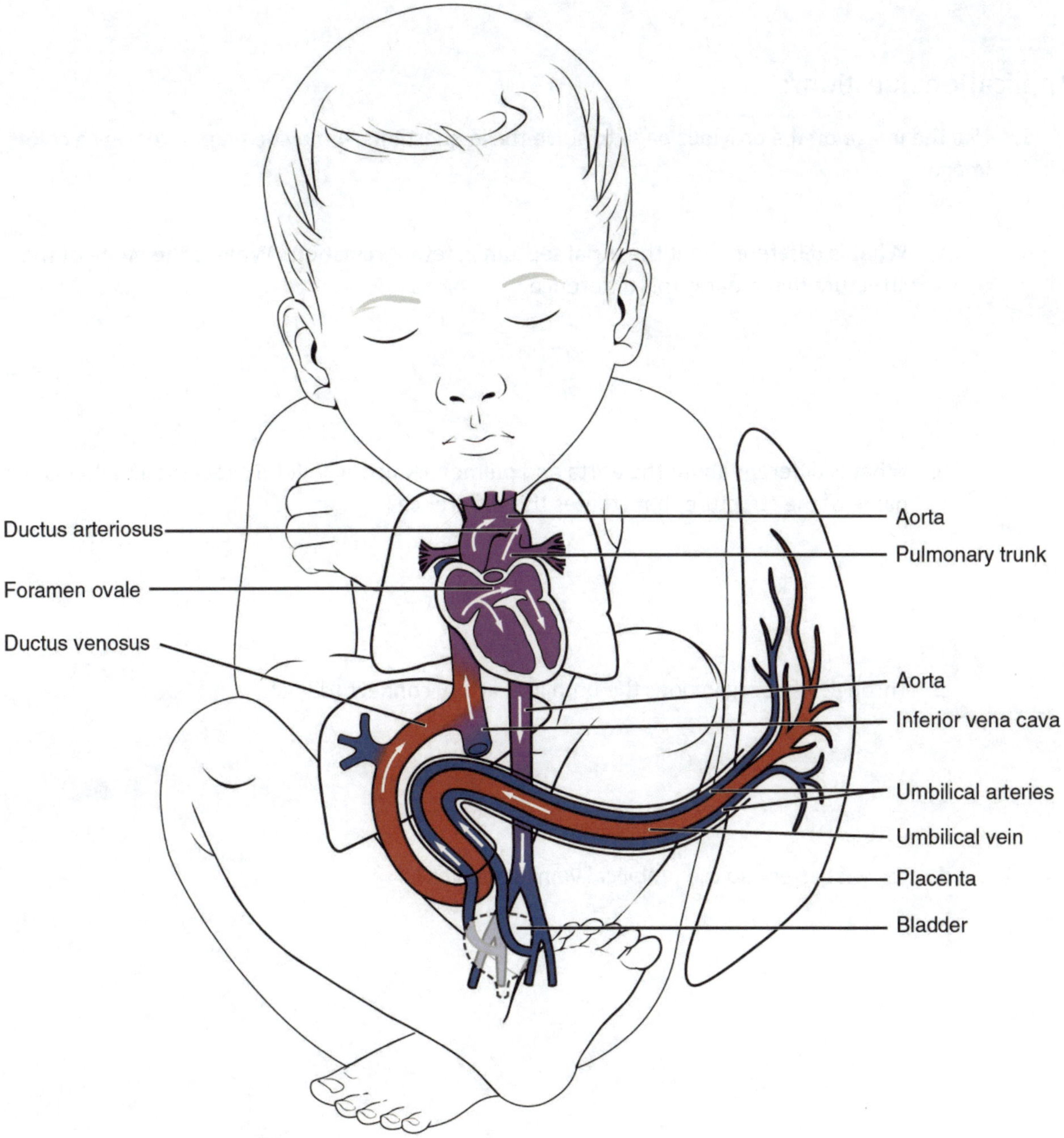


[3]

[3] "Fetal Circulation" by OpenStax. CC BY 3.0 Unported.

## Review questions:

1. How many umbilical veins are there? How many umbilical arteries?

## Application questions:



1. Use the image on the previous page to solve these questions. It may be helpful to view a color image.

    a. What is different about the atrial septum in fetal circulation? What is the name of the structure that creates this difference?

    b. What is different about the aorta and pulmonary artery in fetal circulation? What is the name of the structure that creates this difference?

    c. Through what organ does the umbilical artery connect to?

    d. To what artery do the umbilical veins attach to?

## Exercise 3: Fetal skull

As we learned about earlier in the course, there are two main forms of ossification: endochondral and intramembranous. Here you will learn the structures of the fetal skull before it has completed intramembranous ossification.

**Label the following:**

- Frontal bones (2)
- Parietal bone
- Temporal bone
- Occipital bone
- Sphenoid fontanelle
- Mastoid fontanelle
- Anterior fontanelle
- Posterior fontanelle
- Frontal suture

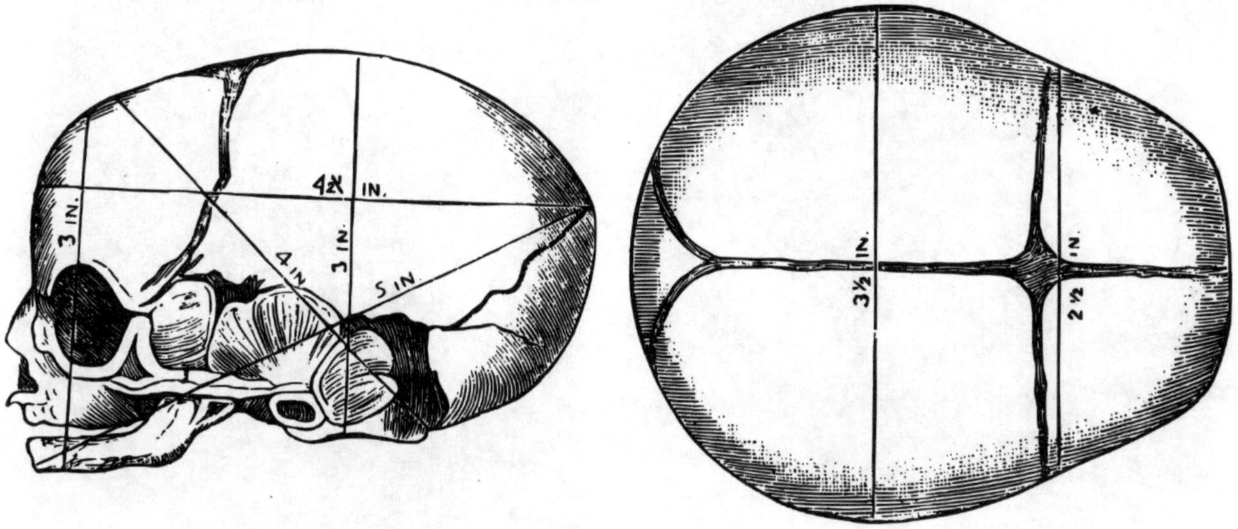


[4] Haultain and Ferguson. Public Domain.

# <u>Exercise 4:</u> Male and Female Reproductive Internal Organs

**Label the following and identify whether it is a male (M), female (F), or both (B) structure:**

- Anus (M/F/B)
- Bladder (M/F/B)
- Cervix (M/F/B)
- Clitoris (M/F/B)
- Corpus cavernosum (M/F/B)
- Ejaculatory duct (M/F/B)
- Epididymis (M/F/B)
- Fallopian tube (M/F/B)
- Fimbriae (M/F/B)
- Foreskin (M/F/B)
- Fornix (M/F/B)
- Labua major and minora (M/F/B)
- Ovary (M/F/B)
- Prostate gland (M/F/B)
- Rectum (M/F/B)
- Seminal vesicle (M/F/B)
- Testes (M/F/B)
- Urethra (M/F/B)
- Uterus (M/F/B)
- Vagina (M/F/B)
- Vertebral column (M/F/B)



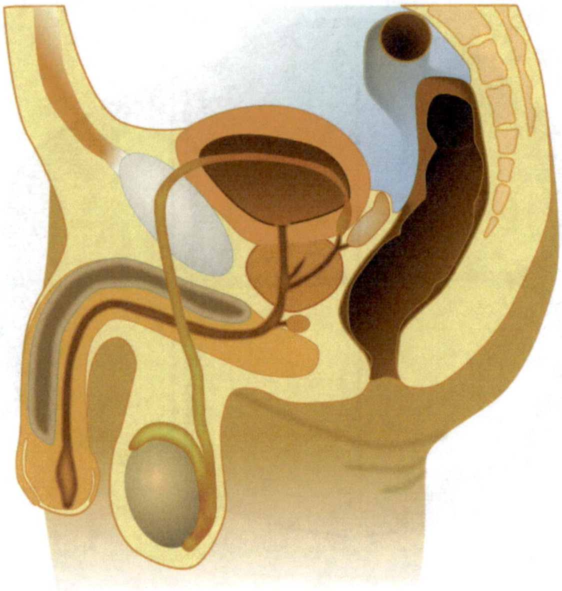

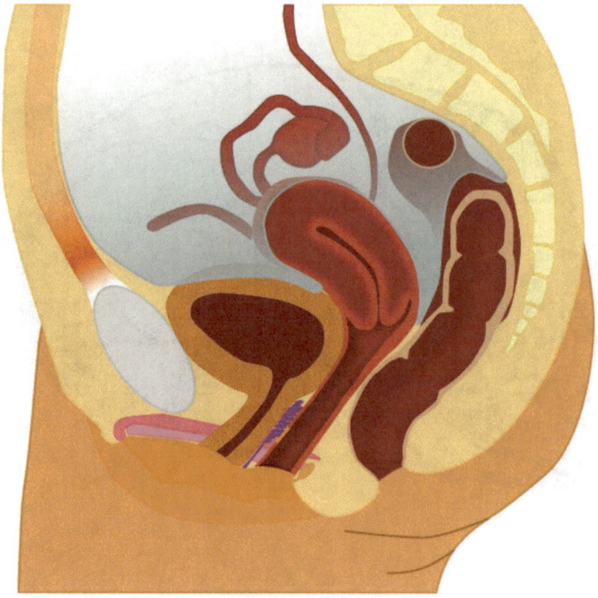

[5] "Lateral anatomy view of the male (left) and female (right) reproductive systems." By Tsaitgaist . CC BY-SA 3.0 Unported